AF788601

Protein Synthesis: Methods and Protocols

Protein Synthesis: Methods and Protocols

Edited by:

Bhupendra Pushkar

Delve
PUBLISHING

www.delvepublishing.com

Protein Synthesis: Methods and Protocols

Bhupendra Pushkar

Delve Publishing
224 Shoreacres Road
Burlington, ON L7L 2H2
Canada
www.delvepublishing.com

Email: orders@arclereducation.com

© 2021 Delve Publishing

ISBN: 978-1-77407-703-0 (Hardcover)

This book contains information obtained from highly regarded resources. Reprinted material sources are indicated and copyright remains with the original owners. Copyright for images and other graphics remains with the original owners as indicated. A Wide variety of references are listed. Reasonable efforts have been made to publish reliable data. Authors or Editors or Publishers are not responsible for the accuracy of the information in the published chapters or consequences of their use. The publisher assumes no responsibility for any damage or grievance to the persons or property arising out of the use of any materials, instructions, methods or thoughts in the book. The authors or editors and the publisher have attempted to trace the copyright holders of all material reproduced in this publication and apologize to copyright holders if permission has not been obtained. If any copyright holder has not been acknowledged, please write to us so we may rectify.

Notice: Registered trademark of products or corporate names are used only for explanation and identification without intent of infringement.

Delve Publishing publishes wide variety of books and eBooks. For more information about Delve Publishing and its products, visit our website at www.delvepublishing.com.

ABOUT THE EDITOR

Dr. Bhupendra Pushkar was born in Mumbai, India. He persuaded his B.Sc. in Microbiology in 2000. He has obtained his M.Sc. degree in Biotechnology in 2002 from University of Mumbai. He pursued his further studies and completed his Doctorate degree in 2010. His area of research broadly includes environmental pollution remediation using biological science. He is currently working on heavy metal bioremediation, PAH bioremediation, plastic degradation, plant tissue culture etc. He is actively involved in the bioremediation research work and trying to develop system for river water cleaning. Since 2008 he is working as assistant professor at University of Mumbai, India. He has published various papers in national as well as international journals and currently guiding three research students. He is Co-Editor in Chief of a journal and reviewer in many journals.

TABLE OF CONTENTS

GLOSSARY OF TERMS

A

Aberrantly - departing from the right, normal, or usual course

Accumulation – a mass or quantity of something that has gradually gathered or been acquired

Aliphatic – relating to or denoting organic compounds in which carbon atoms form open chains

Aminoacyl tRNA Synthetase - Enzyme that "charges" tRNA molecules by catalyzing a bond between the tRNA and a corresponding amino acid.

Antibiotic – a medicine (such as penicillin or its derivatives) that inhibits the growth of or destroys microorganisms

Anticodon – a sequence of three nucleotides forming a unit of genetic code in a transfer RNA molecule, corresponding to a complementary codon in messenger RNA

Arabidopsis – is a genus in the family Brassicaceae. They are small flowering plants related to cabbage and mustard

Archaeology - the study of human activity through the recovery and analysis of material culture

Arsenal – an array of resources available for a certain purpose

Avant-Garde - people or works that are experimental, radical, or unorthodox with respect to art, culture, or society. It is frequently characterized by aesthetic innovation and initial unacceptability

B

Bacteriophage - a virus which parasitizes a bacterium by infecting it and reproducing inside it. Bacteriophages are much used in genetic research

Binding- the action of fastening or holding together, or of being linked by chemical bonds

Biosynthesis – the production of complex molecules within living organisms or cells

Bulkiness - The characteristic of being bulky, or of having bulk.

C

Catalysts – a substance that increases the rate of a chemical reaction without itself undergoing any permanent chemical change

Coaxed - Gently and persistently persuade (someone) to do something.

Cognate - related; connected

Collimated - Made accurately parallel.

Competent – having the necessary ability, knowledge, or skill to do something successfully

Complementarity – a relationship or situation in which two or more different things improve or emphasize each other's qualities

Complex – consisting of many different and connected parts

Concurrent – existing, happening, or done at the same time

Condensation - the change of the physical state of matter from the gas phase into the liquid phase, and is the reverse of vaporization

Conformation – the shape or structure of something, especially an animal

Continuum – a continuous sequence in which adjacent elements are not perceptibly different from each other, but the extremes are quite distinct

Cytochemistry - the branch of cell biology dealing with the detection of cell constituents by means of biochemical analysis and visualization techniques

Cytoplasmic - the cell substance between the cell membrane and the nucleus, containing the cytosol, organelles, cytoskeleton, and various particles

D

Deinococcus radiodurans - is an extremophilic bacterium, one of the most radiation-resistant organisms known

Deoxyribonucleic Acid – DNA, or the double-stranded and very stable molecule that contains the genetic information of most life on earth in the form of variable sequences of 4 different base-pairs.

Discrimination - The unjust or prejudicial treatment of different categories of people, especially on the grounds of race, age, or sex

Drug – a medicine other substance which has a physiological effect when ingested or otherwise introduced into the body

E

E.coli - *Escherichia coli*, also known as *E.coli*, is a Gram-negative, facultative anaerobic, rod-shaped, coliform bacterium of the genus

Electron Microscopy - a technique for obtaining high resolution images of biological and non-biological specimens

Electrostatic Repulsion - Electrostatic repulsion is the result of interaction between the electrical double layers surrounding particles or droplet.

Elongation - the action or process of lengthening something

Embryogenesis – the formation and development of an embryo

Ensemble - Together

Entropically – A measure of the amount of energy in a physical system not available to do work

Evolution – the process by which different kinds of living organisms are believed to have developed from earlier forms during the history of the earth

F

Fate - the development of events outside a person's control, regarded as predetermined by a supernatural power

Fidelity – faithfulness to a person, cause, or belief, demonstrated by continuing loyalty and support

Fruition - the realization or fulfillment of a plan or project

G

Genomes – the haploid set of chromosomes in a gamete or microorganism, or in each cell of a multicellular organism

Goniometer - A goniometer is an instrument that either measures an angle or allows an object to be rotated to a precise angular position. The term goniometry derives from two Greek words, gōnia, meaning angle, and metron, meaning measure.

GTPases - a large family of hydrolase enzymes that bind to the nucleotide guanosine triphosphate and hydrolyze it to guanosine diphosphate

H

Haloarcula marismortui - a halophilic red Archaeon (from the Halobacteriaceae family) found in the Dead Sea, a high saline, low oxygen solubility, and high light intensity environment

Helices – an object having a three-dimensional shape like that of a wire wound uniformly in a single layer around a cylinder or cone, as in a corkscrew or spiral staircase

Homogenous - of the same kind; alike

I

Idiosyncratic - relating to idiosyncrasy; peculiar or individual

Implicated - show (someone) to be involved in a crime

Incompetent - not having or showing the necessary skills to do something successfully

Indispensable - absolutely necessary

Induce – bring on (the birth of a baby) artificially, typically by the use of drugs

Infections – the process of infecting or the state of being infected

Initiation - the action of beginning something

Initiator tRNA - In prokaryotes, called tRNAMetf ; in eukaryotes, called $tRNA_i$; a tRNA that interacts with a start codon, binds directly to the ribosome P site, and links to a special methionine to begin a polypeptide chain.

Intrinsically – in an essential or natural way

K

Kozak's rules - Determines the correct initiation AUG in a eukaryotic mRNA; the following consensus sequence must appear around the AUG: 5'-GCC (**purine**) CCAUGG-3'; the bolded bases are most important.

M

Malignant – (of a disease) very virulent or infectious

Mechanism – a natural or established process by which something takes place or is brought about

Metastable - stable provided it is subjected to no more than small disturbances

Microscopy - the technical field of using microscopes to view objects and areas of objects that cannot be seen with the naked eye

Mitochondria - an organelle found in large numbers in most cells, in which the biochemical processes of respiration and energy production occur

Moiety - In organic chemistry, a moiety is a part of a molecule which is typically given a name as it can be found within other kinds of molecules as well

Molecule - a group of atoms bonded together, representing the smallest fundamental unit of a chemical compound that can take part in a chemical reaction

Monocistronic – is a term used in biochemistry to describe the capacity of eukaryotes to code one gene per one mRNA

Monomeric - is a molecule that forms the basic unit for polymers, which are the building blocks of proteins

Monophyletically - relating to a group of taxa that includes the common ancestor of all the members as well as all descendants of that ancestor

Multifarious – many and of various types

Mutation - the changing of the structure of a gene, resulting in a variant form that may be transmitted to subsequent generations, caused by the alteration of single base units in DNA, or the deletion, insertion, or rearrangement of larger sections of genes or chromosomes

N

Nicking – make a nick or nicks in

Novel – interestingly new or unusual

O

Oligomers – a polymer whose molecules consist of relatively few repeating units

Operons - in genetics, an operon is a functioning unit of DNA containing a cluster of genes under the control of a single promoter

P

Paradoxical - seemingly absurd or self-contradictory

Pathogens – a bacterium, virus, or other microorganism that can cause disease

Peptide Bonds - an amide type of covalent chemical bond linking two consecutive alpha-amino acids from C1 of one alpha-amino acid and N2 of another along a peptide or protein chain

Peptidyl Transferase - RNA-based enzyme that is integrated into the 50S ribosomal subunit and catalyzes the formation of peptide bonds.

Periodate – a salt or ester of periodic acid

Perturbations - anxiety; mental uneasiness

Phenotype - the set of observable characteristics of an individual resulting from the interaction of its genotype with the environment

Polycistronic – a type of messenger RNA that can encode more than one polypeptide separately within the same RNA molecule

Polypeptide - a linear organic polymer consisting of a large number of amino-acid residues bonded together in a chain, forming part of (or the whole of) a protein molecule.

Polyphenylalanine - polypeptide (fragment) containing only phenylalanine

Polysome - mRNA molecule simultaneously being translated by many ribosomes all going in the same direction.

Polyuridylic - RNA or a segment of RNA that is composed of a polynucleotide chain consisting entirely of uracil-containing nucleotides and that codes for a polypeptide chain consisting of phenylalanine residues when functioning as messenger RNA in protein synthesis

Precisely – in exact terms; without vagueness

Priming – a substance that prepares something for use or action

Progenotes - primordial organism in which the relationship between genotype and phenotype was still evolving

Protein -a distinct sequence of nucleotides forming part of a chromosome, the order of which determines the order of monomers in a polypeptide or nucleic acid molecule which a cell (or virus) may synthesize

Protein Synthesis - is a core biological process, occurring inside cells, balancing the loss of cellular proteins through the production of new proteins

Pyrophosphate – a salt or ester of pyrophosphoric acid

R

Redundant - not or no longer needed or useful; superfluous

Resistance – the ability not to be affected by something, especially adversely

Ribonucleic Acid – Otherwise known as RNA, this molecule usually exists as a single-stranded carrier of genetic information.

Ribosomal RNA – RNA molecules associated with ribosomes, some of which are ribozymes and catalyze reactions.

Ribozyme – RNA that acts as a biological catalyst, which in a ribosome helps form peptide bonds.

RNase - ribonuclease is commonly abbreviated as RNase

S

Sequence - the order in which amino-acid or nucleotide residues are arranged in a protein, DNA, etc.

Sequencing – arrange in a particular order

Shine-Dalgarno Sequence - (AGGAGG); initiates prokaryotic translation by interacting with rRNA molecules comprising the 30S ribosome.

Signal Sequence - Short tail of amino acids that directs a protein to a specific cellular compartment.

Start Codon - AUG (or rarely, GUG) on an mRNA from which translation begins; always specifies methionine.

Stringency – the fact or condition of being rigorous and unsparing

Sub-saturating –the condition of being full to or beyond satisfaction

Subsequent -coming after something in time; following

Substantially - to a great or significant extent

Superpose - place (something) on or above something else, especially so that they coincide

Synthesis - the production of chemical compounds by reaction from simpler materials

T

Tandem -having two things arranged one in front of the other

Target – select as an object of attention or attack

Termination - an ending or result of a specified kind

Terminus - A final point in space or time; an end or extremity.

Ternary- using three as a base

Tetrahedral – molecule, there is one central atom bonded to four surrounding atoms, with no lone electron pairs

Toto – totally, entirely

Translation - in molecular biology and genetics, translation is the process in which ribosomes in the cytoplasm or ER synthesize proteins after the process of transcription of DNA to RNA in the cell's nucleus

Translation - the process by which a sequence of nucleotide triplets in a messenger RNA molecule gives rise to a specific sequence of amino acids during synthesis of a polypeptide or protein

Translational Control - the control of protein synthesis by regulation of the translation step, for example by selective usage of preformed mRNA or instability of the mRNA

Treatment – medical care given to a patient for an illness or injury

Truncation - is a searching technique used in databases in which a word ending is replaced by a symbol

U

Ubiquitously – existing or being everywhere, especially at the same time

V

Vertebrates - A vertebrate is an animal with a spinal cord surrounded by cartilage or bone.

LIST OF FIGURES

LIST OF TABLES

ABBREVIATION

AARS	Aminoacyl-tRNA Synthetase
Aa-tRNA	Aminoacyl tRNA
AATSs	Aminoacyl-tRNA Synthetases
AMP	Adenosine MonoPhosphate
AsnRS	Asparaginyl-tRNA Synthetases
ATP	Adenosine Triphosphate
CBP	Cap-Binding Protein
CFPS	Cell-Free Protein Synthesis
ChIPS	Characterization of Inhibitors of Protein Synthesis
CTT	C-Terminal Tail
DC	Decoding Centre
DC	Dendritic Cells
DNA	Deoxyribonucleic acid
EIFs	Eukaryotic Initiation Factors
EM	Electron Microscopy
ES	Expansion Segments
fMet	formylated METhionine
GAV	GlyAla-Val that is glycyl-alanyl-valine
GluRS	Glutamyl-tRNA Synthetases
GTP	Guanosine Triphosphate
HGT	Horizontal Gene Transfer
IC	Interface Canyon
ILL	Institut Laue-Langevin
IRES	Internal Ribosome Entry Site
ITAFs	IRES Trans-Acting Factors
LUCA	Last Universal Common Ancestor
Met-tRNAMet	Methionine
MGH	Massachusetts General Hospital

MLSKB	Macrolides, Lincosamides, Streptogramins B, Ketolides
mRNA	messenger RNA
NMR	Nuclear Magnetic Resonance
NPET	Nascent Peptide Exit Tunnel
NTT	N-Terminal Tail
PABP	Poly-A- Binding Protein
PDB	Protein Data Bank
PrAMPs	Proline-rich Antimicrobial Peptides
PTC	Phenylthiocarbamide
PURE	Protein Synthesis Using Recombinant Elements
RNA	Ribonucleic acid
RNase	Ribonuclease
rRNA	ribosomal RNA
SC Fold	Stem Contact Fold
SEM	Scanning Electron Microscopy
sRNA	Soluble RNA
TEM	Transmission Electron Microscopy
tRNA	Transfer Ribonucleic Acid
uORF	upstream Open Reading Frames

PREFACE

Proteins form an essential component of the structure of various organisms. They are also an important contributor to the nutrition content that the human body needs. The protein is naturally found in many forms and structures in various creatures around the globe and is also manufactured artificially with the help of other protein strands so that better strands can be prepared to provide nutritional and medical support to the humans.

The knowledge of proteins is very limited and is restricted to a handful of books. It is difficult to find relatable content on various kinds of proteins. It requires immense research to reach resources. However, in this book, we have tried to bring the maximum possible amount of information on protein synthesis in a single book so that the readers feel convenient to have access to such knowledge.

Protein synthesis is an important part of the food industry as well as the pharmaceutical industry, which makes the study of such a subject even more important and relevant. This book aims to cover the majority of the basics related to the synthesis of proteins.

The book starts by making the readers aware of the basic structure of the protein and how it exists. The readers are informed about the constitution of proteins, different kinds of their structures and similar introductory things.

The second chapter then focuses on the history of protein synthesis and explains how it has evolved over the years. Then the third chapter brings the attention of the readers to the synthesis of tRNA and the related aspects of it.

The fourth chapter takes the readers through the concepts of polypeptide chain initiation and then elaborates on the elongation of the chain and the way the chain terminates. The fifth chapter talks about them RNA and throws light on its translation to other forms.

The sixth chapter brings the structure and function of the ribosome to the light of readers and also informs the readers about the techniques that are employed to study ribosomes. The seventh chapter takes the readers through the various kinds of resistances that come up in protein synthesis and the inhibitors in the process. The last chapter talks about how the apparatus for the synthesis of proteins have evolved over the years to provide better and improved results in current times with cutting edge technology and developments. I hope that this book proves sufficient in terms of knowledge and information that the readers are pursuing and that it can help them in their research and development purposes.

1

Primary Structure and Three-Dimensional Arrangement of Proteins

CONTENTS

The chapter covers the components of proteins and about the protein families. This chapter describes the primary structure of the protein and also about the primary structure of insulin. This chapter further provides an insights about the three- dimensional structure of proteins that includes the secondary, tertiary structure and the quaternary structure and the symmetry. The chapter majorly focuses on the sequence of amino acids in proteins and the polypeptide chains as well as the bonding. This chapter highlights the knowledge about the structure of the protein.

1.1. INTRODUCTION

The term 'protein' is derived from the '**proteins**'. This is a Greek word which means primary. As per the name represents, the proteins are considered to have paramount importance for biological systems. Three fourth of the total dry body weight is made up of proteins. The proteins are used to build the body. This is because of the fact that all the major structural and functional aspects of the body are carried out by the molecules of proteins.

When there is an abnormality in the protein structure then this will lead to the molecular diseases having profound alterations in the metabolic functions. The proteins tend to contain many major components named Carbon, Oxygen, Hydrogen and Nitrogen and there are some minor components also named Sulfur and Phosphorus.

Nitrogen is said to be among the characteristic of proteins. On an average, in ordinary proteins, the nitrogen content is around 16 % by weight. All the proteins are considered to be the polymers of amino acids.

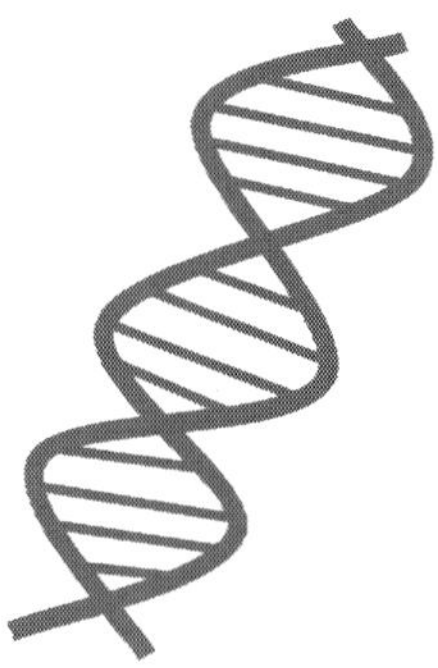

Figure 1.1. DNA - Deoxyribonucleic acid

Source: Image by Free SVG

The proteins play the role of the end products of the decoding process which starts having the information in cellular DNA. Proteins are considered as the workhorses of the cell and therefore, compose the structural and motor elements in the cell.

Furthermore, they serve as catalysts for virtually every biochemical reaction that tends to occur in living things. This incredible array of function is derived from a startlingly simple code which further specifies a hugely diverse set of structures.

In fact, in cellular DNA, each gene tends to contain the code for any particular protein structure. These proteins are assembled with different amino acid sequences. Along with this, they also are combined together by many different bonds and folded into various three-dimensional structures. The folded shape or conformation relies on the linear amino acid sequence of the protein directly.

The shape of a protein is critical to its function. The reason behind this is that it determines whether the proteins can interact with some other molecules. The protein structures are considered to be very complex and therefore, the researchers have gained the ability to easily as well as quickly determine the structure of the complete proteins down to the atomic level.

There are techniques that are used in the period of the 1950s but now in today's era, they were very slow as well as laborious to be used. And this is why the complete protein structures were very slow to be solved.

Conceptually, the early structural biochemists divided the protein structures into four levels to make it easier to get a handle on the complexity of the entire structures. It is important to understand these four levels of protein structure named as primary, secondary, tertiary and quaternary in order to determine the way in which the protein gets its final shape or conformation.

The primary structure of a protein is the unique sequence of the amino acids in every polypeptide chain which makes up the protein. This is just several amino acids that appear in the order in a polypeptide chain and it is not really a structure.

This is referred to as the primary structure of the polypeptide chain because of the fact that the final protein structure ultimately depends on this sequence. An example can be the pancreatic hormone insulin which has two polypeptide chains namely A and B.

1.1.1. What Are Proteins Made Of?

Amino acids are considered as the building blocks of proteins. These are small organic molecules that contain an alpha (central) carbon atom which is linked to an amino group, a hydrogen atom, a carboxyl group and a variable component which is referred to as a side chain.

In a protein, the amino acids are connected along with peptide bonds. This in turn forms a long chain. The formation of the peptide bonds is a biochemical reaction which involves release of water molecule. This is because it joins the amino group of one amino acid to the carboxyl group of neighboring amino acid. The linear sequence of amino acids is considered to be the primary structure of the protein.

The building of proteins is done by using a set of only 20 amino acids and each of them has a unique side chain. There are different chemistries of the side chains of amino acids. There is the largest group of amino acids that have non-polar side chains. There are several other amino acids that have side chains having positive or negative charges, whereas there are others that have polar but uncharged side chains.

The chemistry of amino acids side chains is important for the protein structure and the reason is that these side chains can bond with one another to hold a length of protein in a particular shape or conformation. The ionic bonds can be formed by the charged amino acid side chain and polar amino acids are capable of forming hydrogen bonds. The interaction between hydrophobic side chains takes place with each other by weak van der Waals interactions.

By these side chains, there is a vast majority of non- covalent bonds formation. Furthermore, the cysteines are considered to be the only amino acids which are capable of forming covalent bonds that they do along with their specific side chains. The sequence and location of amino acids in some of the specific proteins tend to guide the occurrence of the bends and folds in that protein and this is because of the side-chain interactions.

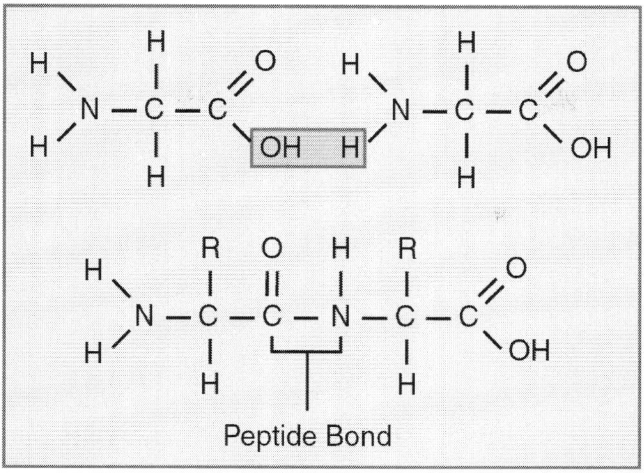

Figure 1.2. Peptide Bonds

Source: Image by Wikimedia Commons

The amino acid sequence is referred to as the primary structure of a protein which drives the folding and intra-molecular bonding of the linear amino acid chain. This eventually determines the unique three - dimensional shapes of the protein. In the neighboring regions of the protein chain, the hydrogen bonding between amino groups as well as carboxyl groups sometimes causes some specified patterns of folding to occur.

These stable folding patterns are known as alpha helices and beta sheets, which make up the secondary structure of a protein. Most of the proteins consist of several helices and sheets, along with some other less common patterns.

The ensemble of formations and folds in a single linear chain of amino acids is sometimes referred to as a polypeptide. And this contains the tertiary structure of a protein. Ultimately, the quaternary structure of a protein refers to those macromolecules which have multiple polypeptide chains or subunits.

The shape that is finally adopted by a newly synthesized protein is typically the most energetically favorite one. As the folding of proteins takes place, they tend to test a variety of conformations before reaching their final form which is said to be unique and compact. The stabilization of the folded proteins takes place by a large number of non – covalent bonds between amino acids.

Along with this, there are chemical forces also in between a protein and its immediate environment which contribute to the protein shape and stability. Let's take an example of the proteins which are dissolved in the cell cytoplasm. These proteins have hydrophilic (water-loving) chemical groups on their surfaces whereas, on the other hand, their hydrophobic (water-averse) elements is tucked inside.

Contrastingly, there are proteins that are inserted into the cell membranes and these proteins display some hydrophobic chemical groups on their surface, particularly in those regions or areas where the protein surface is exposed to membrane lipids. However, it is important to consider that the folded proteins are not frozen into shape. Instead of this, there are atoms within these proteins that have the capability to make small movements.

Although the proteins are considered as macromolecules, they are too small to be visualized, even by using a microscope. Therefore,scientists are required to use some of the indirect methods to figure out what they look like and their folding process.

There is a most commonly used method for studying the protein structures and that is named X-ray crystallography. By using this method, the solid crystals of purified proteins are placed in an X-ray beam and also, there is a pattern of deflected X-rays which is used to predict the positions of the thousands of atoms present in the protein crystal.

1.1.2. How Do Proteins Arrive at Their Final Shapes?

Theoretically, when the connections between their constituent amino acids are formed together, the proteins tend to attain their final shapes without any energy input. However, the cytoplasm is a crowded place. There are many other molecules that are present in the cytoplasm and they have the capability of interacting with a partially folded protein.

There can be irrelevant connections with some of the nearby proteins which can interfere with proper folding and they can cause large aggregates of proteins to form in cells. Therefore, these cells depend on chaperone proteins to prevent these irrelevant connections with some unintended folding partners.

During the folding process, the chaperone proteins surround a protein further sequestering the protein until the completion of the folding process. An example can be bacteria. In bacteria, there are several molecules of the chaperone GroEL which form a hollow chamber around the proteins that are involved in the process of folding. Molecules of a second chaperone that is GroES form a lid over the chamber. There are different families of chaperone proteins that are used by eukaryotes even though they function in similar ways.

There is an abundance of chaperone proteins in cells. These chaperones tend to use the energy from ATP in order to bind and release polypeptides when they undergo the process of folding.

Also, the chaperones provide their assistance in the refolding of proteins in cells. The folded proteins are fragile structures and these structures can easily denature of unfolding.

Although, there are a large number of bonds that hold protein together, among them there are many which are non – covalent and fairly weak. Even considering the normal circumstances, there is a portion of all cellular proteins which are unfolded. Whenever there is an increase in body temperature even by only a few degrees, then there can be a significant increase in the rate of unfolding.

When this happens, then it becomes important to repair the existing proteins which use chaperones and so is much more efficient in comparison to synthesizing the new ones. Cells tend to synthesize the additional chaperone proteins while responding to the 'heat shock'.

1.1.3. What Are Protein Families?

All the proteins tend to bind to other molecules to complete their tasks. The functioning of a protein in a précised manner depends on the ways in which the interaction between its exposed surfaces takes place with those molecules. Proteins having related shapes interact with some of the molecules in similar ways and therefore, these proteins are considered as a protein family. Similar functions are performed by the proteins within a particular family within the cell.

From the same family, the proteins have long stretches most of the time of similar amino acid sequences within their primary structures. Through evolution, these stretches have been conserved and are considered to be important to the catalytic function of the protein.

Let's take an example. At the binding sites, the cell receptor proteins tend to contain different amino acid sequences that further receive chemical signals from outside the cell. Despite this, they are more similar in amino acid sequences which further interact with the common intracellular signaling proteins.

There can be several members in the protein families and they likely evolved from ancient gene duplications. These duplications can result in modifications of protein functions and also, expanded the functional repertoire of organisms over the period.

1.2. PRIMARY STRUCTURE

The primary structure is said to be the simplest level of protein structure which is simply the sequence of amino acids in the polypeptide chain. Let's take an example. There are two peptide chains in the hormone insulin namely A and B.

Each chain has its own set of amino acids which are assembled in a specified manner. For instance, the sequence of the A chain begins with the glycine at the N-terminus and further, ends with asparagine at the C-terminus. Also, it is different from the sequence of the B chain.

The DNA of a gene is used in order to determine the sequence of a protein. This encodes the protein or encodes a part of the protein for multi-subunit proteins.

If there is any change in the DNA sequence of the gene then this can result in a change in the amino acid sequence of the protein. Even if there is a change in just one amino acid in the sequence of the proteins, then this can affect the entire structure and function of the protein.

For instance, it can be said that a single amino acids changes are related to the sickle cell anemia. This is an inherited disease that tends to affect the red blood cells.

In sickle cell anemia, there is one of the polypeptide chains which make up hemoglobin, has slight changes in the sequence. This is the protein that carries oxygen in the blood. The glutamic acid is said to be the sixth amino acid of the hemoglobin β chain. This β chain is one of the two types of protein chains which make up the hemoglobin. The glutamic acid is further replaced by a valine.

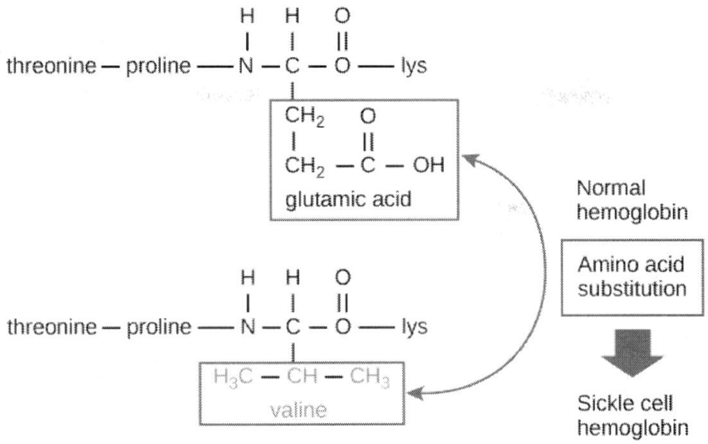

Figure 1.3. Sickle Cell Hemoglobin

Source: Image modified from OpenStax Biology.

Most remarkably, it is important to consider that a hemoglobin molecule is made up of two α chains and two β chains. Each of them contains around 150 amino acids and has a total of about 600 amino acids in the entire protein. There is a difference in between a normal hemoglobin molecule and a sickle cell molecule and that is just 2 amino acids out of the 600 (approximately).

If there is a person whose body makes only sickle cell hemoglobin then his or her body will suffer some symptoms of sickle cell anemia. These symptoms occur because there is a change from the glutamic acid to valine amino acid that makes the hemoglobin molecules assemble into long fibers. The distortion of the disc- shaped red blood cells into crescent shapes occurs due to the fibers.

The sickled cells get stuck when they try to pass through the blood vessels. The stuck cells impair blood flow and can cause some of the serious health problems for the people having sickle cell anemia which include breathlessness, headaches, dizziness and abdominal pain.

1.2.1. Sequence of Amino Acids in Proteins

The protein structures include the primary, secondary, tertiary and quaternary levels. The primary structure represents the number and sequence of amino acids in the protein. The higher levels of organization are decided by the primary structure.

Each polypeptide chain has a unique amino acid sequence that is decided by the genes. The primary structure is maintained which is done by the covalent bonds of the peptide linkages. Whenever there are any changes in the sequence, the peptide is also changed.

1.2.2. Characteristics of a Peptide Bond

The peptide bond is considered to be a partial double bond. The C–N bond is 'trans' in nature. In addition to this, there is no freedom of rotation and the reason is the partial double bond character. The distance is 1.32Å which is in between single bond (1.49Å) and double bond (1.27Å)

The side chains have the freedom of rotation and that can be on either side of the peptide bond. The angles of rotation are referred to as Ramachandran angles. It tends to determine the spatial orientation of the peptide chain. (There was a pioneering work which is done by Dr. G. N. Ramachandran on the structural aspects of proteins during the period of 1950s and 1960s).

1.2.3. Numbering of Amino Acids in Proteins

Considering a polypeptide chain, it can be said that at one end there will be one free alpha- amino group. This particular end is referred to as the amino-terminal (that is N-terminal) end and the amino acid which contributes to the alpha – amino group is known as the first amino acid. Usually, when the sequence of the protein is denoted, then the N-terminal amino acid is written on the left hand side.

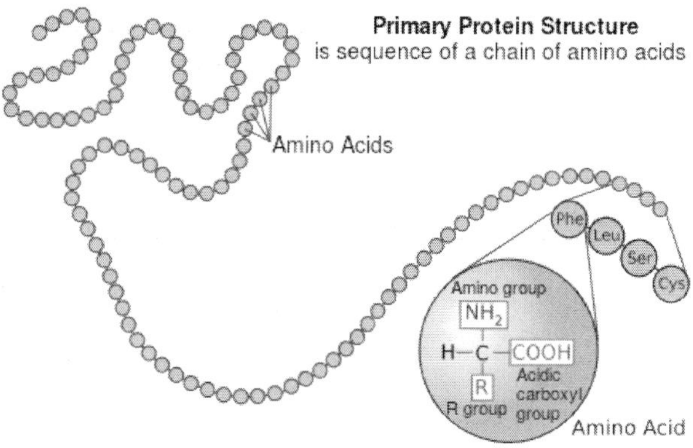

Figure 1.4. Primary Structure of Proteins.

Source: Image by Wikimedia Commons

The starting of the biosynthesis of the protein also takes place from the amino- terminal end. The other end of the polypeptide chain is referred to as the carboxy -terminal end (C-terminal). At this terminal, there is a free alpha carboxyl group that is contributed by the last amino acid.

There is the involvement of all other alpha -amino and alpha carboxyl groups in the formation of the peptide bond. In polypeptides, the amino acid residues are named while changing the suffix "-ine" to "-yl", examples can include Glycine to Glycyl.

NH_2 -Gly-Ala-Val-COOH

In the example provided above, the amino group of glycine has the freedom; but the carboxyl group of glycine is bonded along with the amino group of alanine; the carboxyl group of alanine is bonded with the amino group of valine; on the other hand, the carboxyl group of valine is free. And this is a reason because of which this peptide is named as glycyl-alanyl-valine. The abbreviation thus used is GlyAla-Val, or shown as GAV.

1.2.4. Branched and Circular Proteins

In general terms, the polypeptide chains are said to be linear. Nevertheless, there are branching points in the chains which can further be produced by the interchain disulfide bridges.

There are covalent disulfides bonds that are formed in between different polypeptide chains in the same protein (that is termed as interchain) or in the portions of the same polypeptide chain (that is termed as intrachain) and these bonds are also considered as the part of the primary structure.

It is rare that rather than the alpha COOH group, the gamma carboxyl group of glutamic acid can enter into the formation of the peptide bond formation, for example, Glutathione (which is gammaglutamyl-cysteinyl-glycine).

The term 'pseudopeptide' is used for denoting such a peptide bond which is further formed by carboxyl group, other than that which is present in the alpha position. The protein may be in a circular form which is very rare and the example can be Gramicidin.

1.2.5. Primary Structure of Insulin

For the primary structure of a protein, an example can be insulin. Sanger originally described this in the year 1955. Sanger received the Nobel Prize in the year 1958. Insulin contains two polypeptide chains. The A chain (which is also known as Glycine chain) contains 21 amino acids and B chain (which is termed as Phenylalanine) contains 30 amino acids. They are all combined together by two inter - chain disulfide bonds.

In A chain, the 7th cysteine and in B chain, the 7th cysteine is connected. In a similar manner, in A chain, 20th cysteine and in B chain, 19th cysteine is all connected. There is one more intrachain disulfide bond that is in between 6th and 11th cysteine residues of A chain.

There is a restriction in the variation in the species to amino acids in positions 8, 9 and 10 in A chain and in C-terminal of B chain. During evolution, there is the conservation of the amino acid sequence to a great extent.

1.2.6. Proinsulin

The insulin is synthesized as pro-hormone by the beta cells of the pancreas. The proinsulin is considered to be a single polypeptide chain having 86 amino acids. In biological terms, the formation of the active insulin (2 chains) is done by removal of the central portion of the proinsulin prior to release. Also, the C-peptide (which further connects peptide) is released into the circulation.

1.2.7. Primary Structure Determines Biological Activity

There is a protein having a particular primary structure that will automatically form its natural three -dimensional shapes. Therefore, the higher levels of organization tend to rely on the primary structure. When there is a single amino acid change in the linear sequence then this can have profound biological effects on the function.

Examples can include HbA (that is the normal hemoglobin). In this, the 6th amino acid in the beta chain is glutamic acid. It gets changed to valine in HbS (that is sickle cell anemia).

1.3. THREE-DIMENSIONAL STRUCTURE OF PROTEINS

The covalent backbone of proteins is said to be made up of hundreds of individual bonds. If there was a possibility of free rotation around even a fraction of such bonds, then the assumptions can be done by the proteins as an almost a large number of three – dimensional structures.

However, each and every protein has a particular chemical or structural function. Furthermore, this strongly suggests that each protein has a unique three – dimensional structure. There is a simple fact and that is the proteins can be crystallized and this provides strong evidence that this is the case.

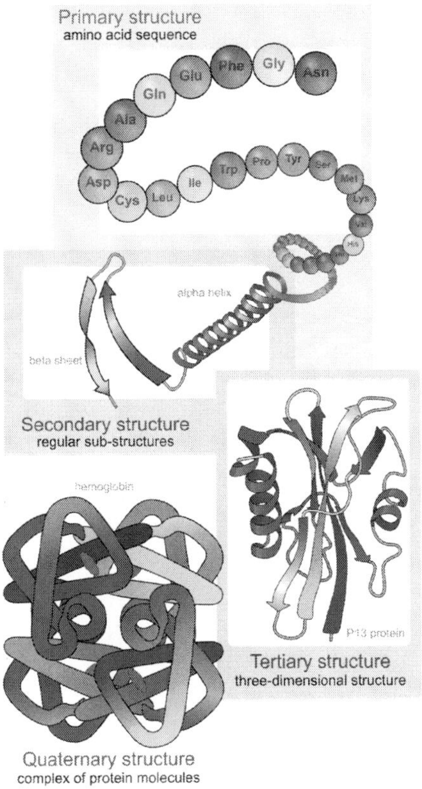

Figure 1.5. Main protein structure levels.

Source: Image by Wikimedia Commons

Generally, in a crystal, the ordered arrays of molecules can form only if the molecular units which make up the crystal are identical. The enzyme urease (which is represented by M_r 483,000) was considered among the first proteins crystallized and this was done by James Sumner in the year 1926).

This was the accomplishment that was demonstrated in a dramatic manner that even large proteins are discrete chemical entities having some kind of unique structures. This further revolutionized thinking about the proteins.

First and foremost, there is a three – dimensional structure of a protein and this is determined by its amino acid sequence. The second point is that the function of a protein tends to depend on its three – dimensional structure. The third point is that the three – dimensional structure of a protein is unique or somewhat similar to that.

The fourth point is that the most important forces which stabilize the specific three – dimensional structure maintained by a provided protein are considered to be non- covalent interactions. Although the structure of proteins is complicated, there are several common patterns that can be recognized.

There is a relationship that is possessed by the amino acid sequence and the three – dimensional structure of a protein. This relationship is an intricate puzzle that has yet to be solved in detail. There are polypeptides that have very different amino acid sequences and these are sometimes assumed to be similar structures. Thus, similar amino acid sequences sometimes can obtain very different structures.

1.4. PROTEIN SECONDARY STRUCTURE

There are different types of secondary structure which are specifically stable as well as they occur in proteins widely. Among them, there are the most prominent ones which are the α helix and β conformations.

Linus Pauling and Robert Corey predicted the existence of these secondary structures in the year 1951. This was the year that was far before the time when the first complete protein structure was elucidated. This was further done by using fundamental chemical principles and a few experimental observations.

While considering secondary structure, it can be said that it is useful so as to classify proteins into two major groups. These groups include fibrous proteins and globular proteins. The fibrous proteins have polypeptide chains which are arranged in long strands or sheets.

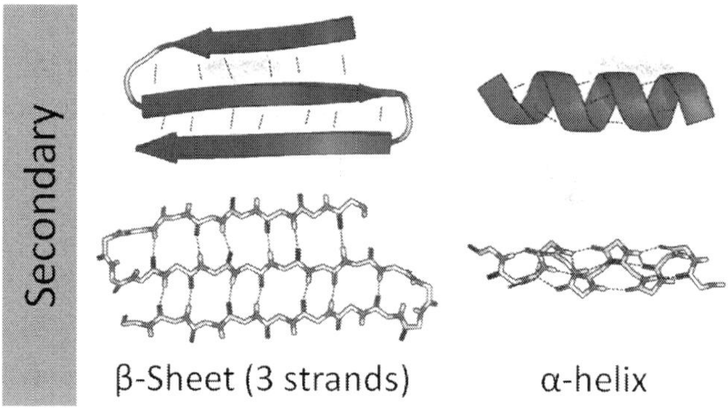

Figure 1.6. Alpha beta structure.

Source: Image by Wikipedia

The globular proteins have polypeptide chains that are folded into a shape that can be either spherical or globular. The fibrous proteins play important as well as structural roles in the anatomy and physiology of vertebrates furthermore; it provides external protection, shape, support and form.

In larger animals, they can contain one – half or more of the total body protein. The globular proteins can include most of the enzymes and peptide hormones. Globular proteins are said to be structurally complex, and the most of the time contain several different types of secondary structure. The fibrous proteins can contain largely of a single type of secondary structure.

This structural simplicity can be a reason because of which some of the fibrous proteins played a key role while developing the modern understanding of protein structure. Moreover, they provide specifically clear examples of the relationship that exists between structure and function.

1.4.1. The Peptide Bond Is Rigid and Planar

Pauling and Corey worked on the protein structure and this was started in the late 1930s. The first of all focused on the structure of the peptide bond. The separation of the carbons of adjacent amino acids can be done by three covalent bonds which are arranged as C_{α}-C-N-C_{α}.

There were many different studies related to the X-ray diffraction which studied the crystals of amino acids and the simple dipeptides and tripeptides. These studies demonstrated the amide C-N bond in a peptide which is shorter in comparison to the C-N bond in a simple amine. Moreover, it shows that the atoms related to the bond are coplanar.

Further indications provided include a resonance or partial sharing of two pairs of electrons in between the carbonyl oxygen and the amide nitrogen. There is a partial negative charge in the oxygen and a partial positive charge in the nitrogen which set up a small electric dipole.

There are four atoms included in the peptide group which tend to lie in a single plane. In this way, the oxygen atom of the carbonyl group and the hydrogen atom of the amide nitrogen are considered to be trans to each other. While considering these studies, the conclusion was further provided by Pauling and Corey that is the amide C-N bonds are not able to rotate in a free manner. The reason behind this is their partial double – bond character.

Therefore, the backbone of a polypeptide chain can be shown as a series of rigid planes that are separated by substituted methylene groups [represented by -CH(R)-]. The rigid peptide bonds tend to restrict the number of conformations which can further be assumed by a polypeptide chain.

The permission of rotation is provided about the N-C_α and the C_α-C bonds. By convention, the bond angles (which are resulted from rotations) are labeled as φ (pronounced as phi) for the N-C_α, bond and ψ (pronounced as psi) for the C_α-C bond. Once again by convention, both φ and ψ are defined as 0° in the conformation. In this, the two peptide bonds which are connected to a single, a carbon are in the same plane.

The φ and ψ can have any value that exists in between -180° and +180°. But on the other hand, there are many values of φ and ψ which are prohibited by steric interference in between atoms in the polypeptide backbone and amino acid side chains. The conformation that is φ and ψ are both 0°, is prohibited because of this. This is used merely as a reference point to describe the angles of rotation.

By the two bond angles including φ and ψ that are repeated at each residue each and every possible secondary structure is described completely. The values which are allowed for φ and ψ can be shown graphically. This can be done by simply plotting φ versus ψ, and this is an arrangement that is referred to as a Ramachandran plot.

The α Helix Is a Common Protein Secondary Structure

The hydrogen bond is important and this is the fact of which Pauling and Corey were aware of. The hydrogen bonds play an important role while orienting polar chemical groups including the -C=O and -N-H groups of the peptide bond. Also, they had the experimental results of William Astbury. William Astbury conducted pioneering X-ray studies of protein in the period of the 1930s.

Further, demonstration provided by Astbury included the protein which makes up hair and wool (this is the fibrous protein α-keratin) has a regular structure which tends to repeat every 0.54 nm. Along with this information and their data related to the peptide bond, and with the help of precisely constructed models, Pauling and Corey further determined the likely conformations of protein molecules.

There can be the simplest arrangement of the polypeptide chain which could assume with its rigid peptide bonds (but there can be other single bonds which are free to rotate) is a helical structure. This was the structure which Pauling and Corey referred to as α helix.

While considering this structure, it can be said that the polypeptide backbone is tightly wound around the long axis of the molecule. Along with this, the R groups of the amino acid residues protrude outward from the helical backbone. The repeating unit is a single turn of the helix, and this further extends approximately 0.56 nm along the long axis which is corresponding closely to the periodicity.

There were observations provided by Astbury on x-ray analysis of hair keratin. The amino acid residues in an α helix have conformations having values of ψ = -45°to -50°and values of φ = -60° and every helical turn includes 3.6 amino acids. Even though a very few left-handed variants have been observed, the twisting of the helix has a right-handed sense in the most common form of the α helix.

Among the two prominent types of secondary structure in proteins, the α helix is considered to be one. It is said to be the predominant structure in α-keratins. Considering the globular proteins, it can be said that around in α helices, one-fourth of all amino acid residues are found, and this fraction tends to vary greatly from one protein to the next.

This can further raise this question that why does such a helix form more readily in comparison to many other possible conformations. The answer to this question can be that it makes optimal use of internal hydrogen bonds. The stabilization of the structure is done by a hydrogen bond in between the hydrogen atom which is attached to the electronegative nitrogen atom of each and every peptide linkage and the electronegative carbonyl oxygen atom of the fourth amino acid on the amino-terminal side of it in the helix.

In such type of hydrogen bonding, each and every peptide bond of the chain tends to participate. Several hydrogen bonds are used to hold each and every successive coil of the α helix to the adjacent coils, which altogether provide considerable stability to the entire structure.

Furthermore, there has been a model – building experiments that have shown that an α helix can form with either L- amino acids or D-amino acids. Nevertheless, all residues are required to be of one stereo-isomeric series. A D-amino acid will disrupt a regular structure that contains L-amino acids and vice versa. The L-amino acids which naturally occur can form either right - handed or left - handed helices. But, there can be rare exceptions where only right - handed helices are found in proteins.

1.4.2. Amino Acid Sequence Affects α Helix Stability

Among all the polypeptides, not each and every one of them can form a stable α helix. There can be additional interactions that can occur in between amino acid side chains. This can further stabilize or destabilize this structure. Let's take an example. If a polypeptide chain has a large number of Glu residues in a long block, then there can be no formation of an α helix at pH 7.0 in this segment of the chain.

The negatively charged carboxyl groups of adjacent Glu residues tend to repel each other strongly. This repulsion is so strong that they overcome the stabilizing influence of hydrogen bonds on the α helix. This can be also the reason, if there are a large number of adjacent Lys and/or Arg residues, having positively charged R groups at pH 7.0, then also, they will repel each other which in turn can prevent the formation of the α helix.

In addition to this, the bulk and shape of some of the R groups can destabilize the α helix or even, they can prevent its formation. Examples can include Asn, Ser, Thr, and Leu residues which tend to prevent the formation of the α helix if they seemed to be occurring closely together in the chain.

When there is a twist of an α helix then it makes sure that the critical interactions occur in between an amino acid side chain and the side chain three (and this can sometimes be four) residues away on either side of it. Most of the time, the amino acids which are positively charged are found three residues away from amino acids which are negatively charged. This permits the formation of an ionic interaction. Similarly, most of the time, two aromatic amino acids are spaced, which further results in a hydrophobic interaction.

There can be a minor constraint on the formation of the α helix and that can be the presence of Pro residues. Considering the proline, it can be said that the nitrogen atom is part of a rigid ring and the rotation about the N-C_α bond is not possible. Additionally, in peptide linkage, the nitrogen atom of a Pro residue has no substituent hydrogen – to – hydrogen bond with other residues. These are the reasons because of which the pro-line is only rarely found within an α helix.

The stability of an α helix can get affected by a final factor and that is the identity of the amino acids. This is located near the ends of the α-helical segment of a polypeptide. In each peptide bond, there exists a small electric dipole. These dipoles tend to add across the hydrogen bonds in the helix so that there can be an increase in the net dipole along with the increase in the helix length.

At either of the ends, the four amino acids of the helix do not participate fully in the helix hydrogen bonds. In actual, the partial positive charges, as well as the negative charges of the helix dipole reside on the peptide amino and carbonyl groups near the amino-terminal and carboxyl-terminal ends of the helix, respectively.

This is the reason because of which most of the time, the negatively charged amino acids are found near the amino terminus of the helical segment. This is the segment where they have a stabilizing interaction along with the positive charge of the helix dipole; at the amino-terminal end, a positively charged amino acid is destabilizing. The contrasting statement is true at the carboxyl-terminal end of the helical segment.

Therefore, it can be said that there are five different kinds of constraints which can affect the stability of an α helix:

- the interaction between amino acids at the ends of the helix and the electric dipole inherent to this structure;
- the electrostatic repulsion (or attraction) between amino acid residues with charged R groups;
- the interactions between amino acid side chains spaced three (or four) residues apart;
- the bulkiness of adjacent R groups; andthe occurrence of Pro residues.

1.4.3. The β Conformation Organizes Polypeptide Chains into Sheets

There is a second type of repetitive structure which was predicted by Pauling and Corey and that is the β conformation. This is said to be the more extended conformation of the polypeptide chains, which can be seen in the silk protein fibroin. The silk protein fibroin is a member of a class of fibrous proteins; β-keratins.

The structure of this has been confirmed by an approach named X-ray analysis. The β conformation is like the α helix and thus, it is common in proteins. In this, the backbone of the polypeptide chain is extended into a zigzag and not like the helical structure.

Considering the fibroin, it can be said that the zigzag polypeptide chains are arranged side by side to form a structure that resembles a series of pleats. This type of structure is known as a β pleated sheet. In the β conformation, the hydrogen bonds can be either intra-chain or inter-chain and that too between the peptide linkages of adjacent polypeptide chains.

Every peptide linkages of β-keratin tends to participate in inter-chain hydrogen bonding. The R groups of adjacent amino acids tend to protrude in contrary directions from the zigzag structure. This in turn creates an alternating pattern similar to the one seen in the side view.

In a β pleated sheet, the adjacent polypeptide chains can be either parallel, while having the same amino-to-carboxyl polypeptide orientation, or anti-parallel, while having the opposite amino-to-carboxyl orientation. Although the repeat period is shorter for the parallel conformation (0.65 nm, as opposed to 0.7 nm for anti-parallel), there is a similarity in the structures.

While considering some of the structural situations, it can be said that there are limitations to the kinds of amino acids which can occur in the β structure. In a protein, when two or more pleated sheets are layered together closely, then the R groups of the amino acid residues on the contact surfaces are required to be relatively small.

β -Keratins includes silk fibroin and the protein of spider webs. These β-keratins contain a very high content of Gly and Ala residues, which are with the smallest R groups. Additionally, in silk fibroin, Gly and Ala tend to alternate over large parts of the sequence.

1.5. TERTIARY STRUCTURE

There is a tertiary structure of a protein which further describes the folding of its secondary structural elements. This tertiary structure also specifies the positions of each atom in the protein. This includes those with its side chains. The protein structures which are known have come been highlighted through X-ray crystallographic or NMR that is nuclear magnetic resonance studies.

In most of these structures, the atomic coordinates are deposited in a Protein Data Bank (PDB) database . Such types of data are available readily on the Internet and this further allows the tertiary structures of various proteins that are to be analyzed as well as compared. There is a common feature in the protein tertiary structure which reveals many things related to the biological functions of the proteins and their evolutionary origins.

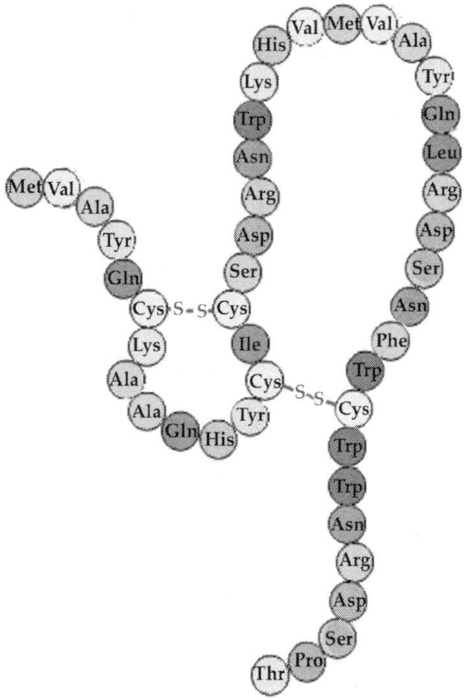

Figure 1.7. Tertiary protein structure.

Source: Image by Wikimedia Commons

1.5.1. Determining Protein Structure

X-ray crystallography is considered to be one of the most powerful methods to study the macromolecular structure. As per the optical principles, the uncertainty that occurs while locating an object is said to be approximately equal to the wavelength of the radiation which is further used to observe it.

A molecule can be directly imaged by using the X-Rays. The reason behind this is that the X-rays wavelengths are comparable to the covalent bond distances (which can be equal to 1.5 Å). Such types of individual molecules cannot be further seen in a light microscope asthe visible light has a minimum wavelength that is equal to 4000 Å.

When the visualization of a crystal of the molecule is done and it is exposed to a collimated (which means parallel) beam of X-rays, then the atoms in the molecule scatter the X-rays. This, in turn, cancels or reinforces the scattered ways in a process which are further referred to as diffraction. The result thus is a diffraction pattern that is recorded on photographic film or by a radiation counter.

Then, the intensities of the diffraction maxima (which refer to the darkness of the spots on the film) are used to mathematically construct the 3-D image of the crystal structure.

Contrastingly, the fibrous proteins do not crystallize. Rather, they can be drawn into fibers. These fibers tend to have X-ray diffraction patterns which contain only a few spots and therefore, they contain little structural information in comparison to others. In a similar way, the diffraction pattern of a DNA fiber is simple while relating it with others. Almost exclusively, the X-rays interact with the electrons in the matter and not the atomic nuclei.

Therefore, an X-ray structure is said to be an image of the electron density of the object that is to be studied. Three – dimensional contour map is used to show the information. The hydrogen atoms have only one electron and therefore, they are not visible in macromolecular X-ray structures. Considering the small organic molecules, it can be said that the x-ray structures can be determined with a resolution on the order of 1 Å.

Among the protein crystals, there are a few which have this degree of organization. Moreover, among all the proteins, not all can be coaxed in order to crystallize or to precipitate in ordered 3-D arrays. The protein crystals do form and they tend to differ from those having most small organic molecules while being highly hydrated. Typically, the protein crystals are 40 % to 60 % water by volume.

There is large solvent content which gives the consistency of soft and jellylike to the protein crystals. Thus, the molecules are disordered by a few angstroms. Furthermore, this tends to limit their resolution to approximately 2 to 3.5 Å, although a few protein crystals are better ordered which means that they have higher resolution.

A resolution of a few angstroms is too coarse to reveal the positions of individual atoms, but the tracing of the distinctive shape of the polypeptide backbone can be done in a usual manner. Therefore, the positions and orientations of its side chains can be deduced further.Nevertheless, there are many side chains have similar sizes and shapes. Therefore, the knowledge of the primary structure of the protein is required to fit the sequence of amino acids to its electron density map. There are certain mathematical techniques that can refine the atomic positions to within 0.1 Å in high – resolution structures.

1.6. QUATERNARY STRUCTURE AND SYMMETRY

Among all the proteins, most of them (specifically those having molecular masses of 100 kD) contain more than one polypeptide chain. There is a particular geometry related to these polypeptide subunits. The spatial arrangement of these subunits is referred to as the quaternary structure of a protein.

There are several reasons for the fact that the multi – subunit proteins are so common. Considering the large assemblies of proteins, like the collagen fibrils, it can be said that there are advantages of subunit construction over the synthesis of one huge peptide chain which are analogous to those further using prefabricated components while constructing a building.

The defaults can further be repaired and this is done by simply replacing the flawed subunit. There can be a difference in the site of a subunit from the site of the assembly into the final product. Only genetic information is required for specifying the whole edifice and that is the information that specifies its few different self – assembling subunits.

Considering the case of enzymes, the increase in the size of protein tends to better fix the three – dimensional positions of its reacting groups. The size of an enzyme is increased by the combination of identical subunits and this is said to be more efficient in comparison to the increasing length of its polypeptide chain. This is because every subunit has an active site. The subunit construction of a number of enzymes tends to provides a structural basis while regulating their activities.

1.6.1. Subunits Usually Associate Non - Covalently

There can be identical or non – identical polypeptide chains that can be contained in a multi – subunit protein. Let's take an example of hemoglobin. Hemoglobin has subunit composition a_2b_2.

There are proteins have more than one subunit and they are referred to as oligomers. The identical units of them are referred to as protomers. Therefore, protomers can contain one polypeptide chain or a number of different, unlike polypeptide chains.

In such a context, the hemoglobin is said to be a dimer of ab protomers. The contact regions that exist in between subunits closely resemble the interior of a single – subunit protein. There are closely packed non – polar side chains contained in them, which have hydrogen bonds that involve the polypeptide backbones and their side chains and interchain disulfide bonds, in some of the cases.

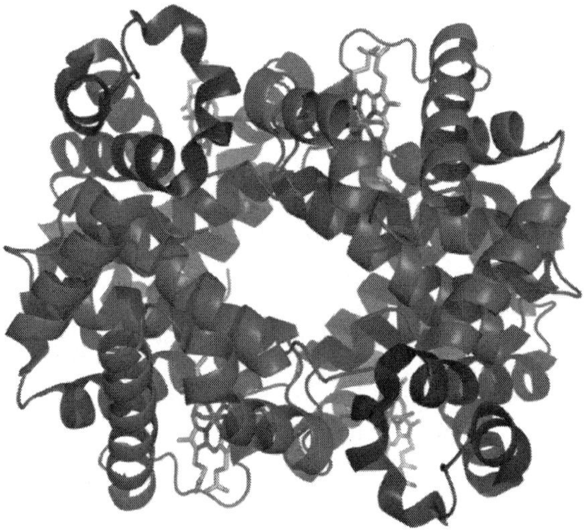

Figure 1.8. 1GZX Hemoglobin.

Source: Image by Wikimedia Commons

1.6.2. Subunits Are Symmetrically Arranged

Considering the vast majority of oligomeric proteins, it can be said that the symmetrical arrangement of the protomers is done. This means that every protomer tends to occupy a geometrically equivalent position in the oligomer. There cannot be inversion or mirror symmetry in the proteins. When the protomers are brought into coincidence then this would require the conversion of the chiral L residues to D residues.

Therefore, the proteins can only have rotational symmetry. Considering the simplest type of rotational symmetry, cyclic symmetry, the protomers are said to be in relation to a single axis of rotation. The objects having two-, three-, or n-fold rotational axes have C_2, C_3, or C_n symmetry, respectively. The C_2 symmetry is said to be the most common; and on the other hand, the higher cyclic symmetries are relatively rare.

Dihedral symmetry (D_n) refers to a more complicated type of rotational symmetry. This symmetry is generated while the intersection of an n-fold rotation axis takes place with a two-fold rotation axis at right angles. An oligomer having D_n symmetry contains 2n protomers.

The D_2 symmetry is said to be the most common type of dihedral symmetry in the proteins. There are some other possible types of rotational symmetry which are those of a tetrahedron, cube, and icosahedrons. Some multi - enzyme complexes, as well as spherical viruses, are built on such geometric plans.

1.7. CONCLUSION

Proteins are said to be built as chains of amino acids. These amino acids then fold into unique three-dimensional shapes. There is bonding within protein molecules which in turn helps in stabilizing their structure. Furthermore, the final folded forms of proteins are very well - adapted for their functions.

Proteins are easy to synthesize in comparison to nucleic acids. Before the RNA world, there likely existed a stage which can be referred to as the protein world. In these proteins, some of the three - dimensional (3-D) peptide structures have been conserved since that time. Moreover, these structures may constitute the oldest biological relics in existence

The conclusion that can be drawn from this chapter is that ten percent of amino acids in typical proteins belong to a nest and therefore, it is a common motif. It is found that a five – residue nest is said to be a part of the well – known P-loop which is a recurring feature of many ATP or GTP – binding proteins. It also has the function of binding the phosphate part of such ligands.

REFERENCES

1. Bioinfo.org.cn. n.d. Chapter 7 : The Three-Dimensional Structure Of Proteins. [online] Available at: <http://www.bioinfo.org.cn/book/biochemistry/chapt07/bio1.htm> [Accessed 6 May 2020].
2. Courses.lumenlearning.com. n.d. Protein Structure | Introduction To Chemistry. [online] Available at: <https://courses.lumenlearning.com/introchem/chapter/protein-structure/> [Accessed 6 May 2020].
3. Khan Academy. n.d. Protein Structure: Primary, Secondary, Tertiary & Quatrenary (Article) | Khan Academy. [online] Available at: <https://www.khanacademy.org/science/biology/macromolecules/proteins-and-amino-acids/a/orders-of-protein-structure> [Accessed 6 May 2020].
4. Milner-White, E., 2019. Protein three-dimensional structures at the origin of life. *Interface Focus*, [online] 9(6), p.20190057. Available at: <https://royalsocietypublishing.org/doi/10.1098/rsfs.2019.0057> [Accessed 6 May 2020].
5. n.d. *PROTEINS: THREE-DIMENSIONAL STRUCTURE*. [ebook] p.4. Available at: <http://www.lacim.uqam.ca/~chauve/Enseignement/BIF7001/H03/Documents/Voet_Voet_chapter6.pdf> [Accessed 6 May 2020].
6. Nature.com. n.d. *Protein Structure | Learn Science At Scitable*. [online] Available at: <https://www.nature.com/scitable/topicpage/protein-structure-14122136/> [Accessed 6 May 2020].
7. Vasudevan, D., S, S. and Vaidyanathan, K., 2016. Proteins: Structure and Function. *Textbook of Biochemistry for Medical Students*, [online] pp.36-36. Available at: <https://www.researchgate.net/publication/327245570_Chapter-04_Proteins_Structure_and_Function> [Accessed 6 May 2020].

2

History of Protein Synthesis (Cell-Free Protein Synthesis)

CONTENTS

Cell-free protein synthesis (CFPS) is also known as *in vitro* protein synthesis. CFPS is basically the use of biological machinery for producing the protein in a system that is cell-free, meaning production without the use of living cells. This chapter is all about the historical marks and development in the cell-free protein synthesis. At the starting of the chapter, the early history of protein synthesis is explained along with its archaeology. The mid part of the chapter tells about the 1950s basic mechanism, landmark advance in the development of cell-free protein synthesis, steps towards an *in-vitro* protein synthesis, the golden age of translation, 1960s. The chapter ends with a brief synopsis of the 1970sand 1990s, translational control and its origin.

2.1. INTRODUCTION

Cell-free protein synthesis (CFPS) is a system that is a simple, sensitive and rapid tool. The Cell-free protein synthesis is devoid of membrane- bound barriers but all the mandatory substrates, machinery and biomolecules are present in the system for synthesis of the desired proteins. The CFPS system has the potential to overcome the loopholes present in the in-vivo production systems.

It has approved to be a promising tool when used in both basic and applied scientific research. CFPS has emerged as a powerful tool in the research of biology which facilitates an organization that is simplified and its way containing the desired experiments and variety of reaction conditions.

The existence of the system has proved useful in the production of toxic and complex proteins, genetic codes, assembly of viruses inmetabolic engineering. The Cell-free protein synthesis is now seen as a potential technology that would help in high throughput production of enzymes, membrane proteins and therapeutics.

In molecular biology, one of the most widely used methods is the Cell-free protein synthesis. Safety protein synthesis basically means making of proteins without cell. However, it usually takes a few hours in the preparation of the proteins using the Cell-free protein synthesis method. While the production of proteins in cells takes many days or maybe some weeks.

If a user starts using the commercial system, the first-time users may obtain newly synthesized proteins on the same day they begin to use the system. Instead of making the proteins in owned laboratories, many users would like it to be used commercially. Basically, it is necessary to understand the processes and mechanisms that are there in the Cell-free protein synthesis.These are reviewed in Spirin and Schwartz, 2008.

Cell-free protein synthesis is a powerful technology platform that is simple and efficient for the production of protein. It is used for meeting the growing demands of its production. After much studies and research being done for understanding the need for transcription and translation, there are some recent advances and development been done on the foundational research tool. The research has made cost- effective microscale to the manufacturing of the complex protein synthesis.

The advances are in terms of the yields that the protein generates. There is an excess of grams protein produced per liter reaction volume of the protein yields. The reactions in batch can last for many hours, the reaction scale to reach a 100-litre milestone and there are reduced costs for orders of magnitude.

The advances in the CFPS system have taken out new applications in the library of protein synthesis. The new application for the structural biology, functional genomics, the expression of virus-like particles, the production of personalized medicines. The coming years promise to have new processes in the cell-free protein synthesis. The short time period to be crucial for the innovative approaches to the different range of applications.

2.1.1. Protein Biosynthesis

Paul Charles Zamecnik (22 November 1912–27 October 2009), an American scientist, often regarded as "the Nestor of protein synthesis research" was once called at the Harvey Society in New York to deliver a lecture in May 1959. Zamecnik chooses to deliver a prestigious lecture on the topic of "Historical and Current aspects of Protein Synthesis" and he traced them back to "careful, patient studies" extending, as the scientist said, "over half a century".

Figure 2.1. Paul Charles Zamecnik, an American scientist regarded as the Nestor of protein synthesis research

Source: Image by Wikimedia Commons

Zamecnik started the lecture by mentioning different people and their researches stating them as an impressive list of pioneers, who all, according to the scientist, "blazed the trail to the present scene".

Starting with Franz Hofmeister, an early protein scientist and Emil Fischer, a German chemist who won the Nobel Prize for Chemistry 1902 & recognized the peptide bond structure of proteins and went on to a Professor of Biochemistry who realized that peptide bond formation was of an endergonic nature, i.e., Henry Borsook.

The author also talked about the Nobel Prize winner of Physiology/ Medicine 1953, i.e., Fritz Lipmann for the suggestion about the participation of a high-energy phosphate intermediate in protein synthesis and added the Jewish-German biochemist, Mr. Max Bergmann's theory who determined the specificity of proteolytic enzymes.

There was discussion about the use of radioactive tracer techniques in following the metabolic pathway, which was developed by Rudolf Schoenheimer and David Rittenberg and the topic like the possible role of RNA in protein synthesis mentioning the names of Torbjörn Caspersson and Jean Brachet, who became aware of the possible role of RNA in protein synthesis.

In the Lecture, also included was the contribution by Frederick Sanger, the winner of Nobel Prize for Chemistry in 1958 and 1980, who unraveled the first primary structure of a protein, showing the specificity and uniqueness of the amino acid composition of insulin. Finally concluded this list of pioneers of protein synthesis by adding the name of George Palade. Palade, who won the Nobel Prize for Physiology/Medicine in 1974 and gave visual evidence for the particulate structures in the cytoplasm acting as the cellular sites of protein synthesis.

Figure 2.2. Frederick Sanger, a Nobel prize winner in chemistry

Source: *Image by Wikimedia Commons*

2.1.2. Early History of Cell-Free Protein Synthesis

At the time when molecular biology was at the beginning of the emergence, many laboratories raised a question on to the reasons for the incorporation of the amino acids into proteins. Using different extracts, different researchers studied the cell-free protein synthesis. Zamecnik et al., (1948), used the extracts of rat liver cells, Gale and Folkes, 1954; Lamborg and Zamecnik, 1960, used bacteria, Bank and Marks, 1966; Schweet et al., 1958, used human and rabbit reticulocytes, Keller, and Littlefield, 1957, used ascites cell and wheat germ was used by Marcus and Feeley, 1966.

It was discovered that there is an occurrence of protein synthesis in the microsome fraction which requires Adenosine Triphosphate (ATP) and Guanosine Triphosphate (GTP) (Hoagland et al., 1956; Littlefield et al., 1955). The correspondence between the encoded nucleotide triplets and the amino acids was found for the first time by Nirenberg and Matthaei in the year 1961. It demonstrated the cell-free synthesis of polyphenylalanine from synthetic Polyuridylic acid in the *E.coli* extract (Nirenberg and Matthaei, 1961).

The following years made a remark in the identification of the amino acids encoded by the remaining triplets. This was made possible through the important role played continuously by the Cell-free synthesis. This basically means breaking the remaining characters in the genetic code (Nirenberg, 2004) and the "central dogma of molecular biology" (Crick, 1970; Strasser, 2006).

2.2. THE ARCHAEOLOGY OF PROTEIN SYNTHESIS – THE 1940S: FORGOTTEN PARADIGMS

The early years of the 1940s were the peak days for Lily Kay as Kay has described that year as the 'protein paradigm of life'. The protein was considered as the key substance by the colleagues of Oswald Avery and Avery at the Rockefeller Institute with DNA. It was not only for the biochemical function but also of hereditary transmission (from Delbrück [41] to Haurowitz [42]).

Surprisingly, there was early attention on protein. However, throughout the 1940s, the mechanism of protein synthesis remained a closed book. During the decade, the thoughts on the mechanism of protein synthesis were centered on the conception. This was being favored by many eminent biochemists of that time such as Max Bergmann and Joseph Fruton. According to them, the mechanism of protein synthesis might be based on a reversal of proteolysis.

However, the concept of proteolysis remained a controversial issue. At that time, it could not be reconciled with the endergonic nature of peptide bond formation. Although it appeared to be evident from the investigations done at the California Institute of Technology in Pasadena. The investigation was carried out by Henry Borsook.

As per the investigations of Henry Borsook, there might be some sort of involvement of the activation of amino acids in the formation of the peptide bonds. This being the case prior to their condensation. This has been already speculated by biochemists, Herman Kalckar and Fritz Lipmann in the early 1940s.

There has been a sophisticated growth in the experimental enzymology and structurally, physical, and chemical analysis of proteins, there are new powerful devices, for example, electrophoresis, X-ray crystallography and chromatography. But the classical biochemistry so far has not provided any definite answer on the question of the cellular mechanism in protein biosynthesis.

On the part of cytochemistry, some observations remained unpredictable but they were interesting. The techniques for the measurement of UV absorption of nucleic acids within cells and the UV microscopy of cells were developed by Torbjörn Caspersson, who was from Stockholm and Jack Schultz from the Kerckhoff Laboratories in Pasadena. This was developed around the year 1940.

By doing this, the correlation growth was made possible i.e., with the increased presence of ribonucleic acids at certain nuclear and cytoplasmic locations in the production of proteins. The same conclusions were drawn byJean Brachet, Raymond Jeener, and Hubert Chantrenne in Brussels. They concluded the same o the basis of differential staining and in situ RNase digestion of tissues.

The particular structure of the cytoplasm was made clear by employing high-speed centrifugation in the year 1930s. It has been derived from other lines of research. In order to characterize and isolate mitochondria, Normand Hoerr and Robert Bensley used centrifugation in Chicago. Albert Claude, who was the winner of the Nobel Prize for Physiology or Medicine in 1974 was working on the isolation of Peyton Rous' chicken sarcoma agent. Claude started working on this in James Murphy's Laboratory at the Rockefeller Institute in New York.

Figure 2.3. Albert Claude, a Nobel Prize winner in Physiology or Medicine, was working on the isolation of Peyton Rous' chicken sarcoma agent

Source: Image by Nobel Prize

There Claude realized that there is the same constitution of chemicals while sedimenting the particles from the infected cells. Claude found that it has exactly the same chemical constitution than those sedimented from normal chick embryo tissue and thus was cellular constituents.

Claude found it to be smaller than the mitochondria and accordingly termed them "microsomes". It was found to be particularly rich in ribonucleic acid (Albert Claude, The Coming Age of the Cell, 1975). Claude always wanted them to be placed in the category of 'Plasmagenes'. However, Walter Schneider and George Hogeboom, both together resisted Claude's attempt of this for correlating specific and unique enzymatic functions with them.

However, in contrast, the preferential objects for ultracentrifugation were the microsomes. The methods of centrifugation however pointed to greatly varying size of the particles and if there is any definable shape of the particles. These methods were found out from the Brachet's laboratory in Brussels and of Cyrus Barnum and Robert Huseby from the Division of Cancer Biology at the University of Minnesota in Minneapolis.

There were no particular experimental efforts were made that reinforces the line of argument given by Brachet, that there is a close connection between microsomes and protein synthesis. However, there were set of procedures for the gentle isolation of cytoplasmic fractions that was the result of the various efforts of an *in vitro* characterization of the cytoplasm through ultracentrifugation. This also proved to be useful while experimenting with other things.

2.3. BASIC MECHANISMS – THE 1950S

Another approach was adopted for assessing the metabolic events and this was expected to be changed between 1945 and 1950. AfterWorld War II, various radioactive tracers were readily available such as ^{35}S, ^{14C}, and ^{3}H as well as ^{32}P. These were used in research for the by-product of expanding reactor technology in a wider scientific constituency.

There was a resultant attack through the use of amino acids on the mechanism of protein synthesis. This was greatly benefitted from the ideal range of resources that were made available for research after the war. The American Atomic Energy Commission made an effort to demonstrate the potentials of peaceful use of radioactivity.

Many kinds of research that were regarding the cancer research programs provided the background to help the protein synthesis research at that time period. The reason for the various researches for the protein synthesis in the decade between 1950 and 1960 was carried based on experimental systems which were derived from higher animals, mainly rat liver, and not on bacteria.

2.4. LANDMARK ADVANCES IN THE DEVELOPMENT OF CELL-FREE PROTEIN SYNTHESIS SYSTEMS

Because of the fact that the preparation processes did not remove mRNA that is messenger RNA, the cell extracts (which contain early cell-free systems) further synthesized native proteins most of the time from the endogenous mRNA. These extracts were considered to be essentially supernatants and they were obtained by centrifugation at high speed of lysates from disrupted cells. An example can include at 30,000g (this is what is referred to as S30 extract) (Lamborg and Zamecnik, 1960; Matthaei and Nirenberg, 1961; Schweet et al., 1958; Zamecnik et al., 1948).

Later, Nirenberg and Matthaei incubated the *E.coli* S30 extract which was done in the presence of 20 amino acids and ATP regeneration components (Nirenberg and Matthaei, 1961) to direct the protein synthesis machinery for translating the exogenous mRNA.

This type of "incubated S30" further improved the amount of the protein which was synthesized from exogenous mRNA. This is done possibly by freeing ribosomes from getting attached to endogenous mRNA (which is referred to as ribosome run-off) and by allowing endogenous mRNA that is to be degraded by ribonucleases in the extract.

In a similar way, Pelham and Jackson added a micrococcal ribonuclease that is Ca^{2+}-dependent while preparing the rabbit reticulocyte extract (that is lysate) in order to degrade endogenous mRNA (Pelham and Jackson, 1976). Subsequently, the micrococcal ribonuclease can be inactivated by EGTA which is done to prevent its digestion of exogenous mRNA.

There is no translation of a significant amount of endogenous proteins that take place from the wheat germ extract. This is the reason why there is no need for incubation or nuclease-treatment (Roberts and Paterson, 1973).
There was the next landmark while improving the cell-free protein synthesis. This was the introduction of coupled transcription-translation systems. These were the systems in which the DNA acted as the template rather than mRNA (Chen and Zubay, 1983; Zubay, 1973). *E.coli* RNA polymerase was used by Zubay and colleagues in the S30 extracts for transcribing the DNA into mRNA for protein synthesis, and majorly for performing the study on the gene regulation *in vitro*.

There were other researchers who later used stronger phage promoters and the more efficient phage RNA polymerases (T7 or SP6) for coupled transcription-translation from the DNA template to obtain higher amounts of mRNA (Craig et al., 1992; Krieg and Melton, 1987; Nevin and Pratt, 1991).

The coupled transcription-translation is done based on the phage promoters and RNA polymerases. This translation has been adopted by most cell-free systems since then. Along with this while using the DNA template which was more stable and avoiding a separate *in vitro* transcription step, the coupled transcription and translation in the system that is the cell-free result in a higher protein synthesis yield most of the time. This is because of the fact that the newly synthesized mRNA is immediately translated into protein and this can further reduce the adverse effects of mRNA degradation or inhibitory structures.

The conduction of the cell - free protein synthesis is commonly done in batch format. Nevertheless, the protein synthesis reactions stop after a short period of time most of the time and there are a number of cases the reactions stop in only 1 hr. or so. This is because of the depletion of ATP and GTP, degradation of mRNA and collection of by – products for inhibiting concentrations.

Normally, the batch cell-free protein synthesis tends to yield μg ml−1 of protein. The introduction of the concept of the continuous cell – free protein synthesis in both *E.coli* as well as eukaryotic cell – free systems is done by Spirin and colleagues (Spirin et al., 1988; Ryabova et al., 1998; Shirokov et al., 2007).

The continuous cell-free protein synthesis tends to circumvent some issues of the batch format and this is done by using a flow chamber or dialysis membrane. This flow chamber or dialysis membrane is used in order to maintain the levels of small molecules energy sources as well as amino acids and helps in removing inhibitory by-products such as phosphate, at the same time. This further resulted in a continuous cell-free protein synthesis reaction which can last more than 10 hours most of the time and yields mg ml^{-1} of protein.

Because of the continuous improvement that is done by academic and industrial researchers, improvements have been done by cell extracts in Cell-free systems significantly. Nirenberg and Matthaei were the first described the "incubated S30" (Nirenberg and Matthaei, 1961). The development of the robust protocols was done by Zubay, Pratt and colleagues in order to make the *E.coli* extract which has become the standards (Pratt, 1984; Zubay, 1973).

There were some other important improvements that further included the use of *E.coli* strains that lacked RNase I (MRE600) and Exonuclease V (A19) in order to stabilize RNA and linear DNA templates, respectively (Gesteland, 1966; Yang et al., 1980).

In recent times, Swartz and colleagues pioneered the use of genome engineered *E.coli* strains. This is done in order to stabilize amino acids during *in vitro* reactions (Michel-Reydellet et al., 2004), and during the *in vitro* activation of endogenous metabolic pathways in order to sustain high - yield protein synthesis (Jewett et al., 2008; Jewett and Swartz, 2004a), as well as the systematic optimization of *E.coli* growth conditions and extraction procedures for the reduction of the production cost (Kim et al., 2006; Liu et al., 2005; Zawada and Swartz, 2005; Zawada and Swartz, 2006).

Since the time of their initial development, there was no change in the protocols in order to make the "classic" eukaryotic extracts, rabbit reticulocyte lysate (Jackson and Hunt, 1983) and wheat germ (Anderson et al., 1983; Erickson and Blobel, 1983). Moreover, the protein synthesis yields from these eukaryotic extracts are generally lower in comparison to the *E.coli* extract. In the recent period, there were many researchers who developed the improved protocols in order to make highly active extracts from wheat germ and mammalian cells.

The remnants of endosperm on the surface of wheat germ, as per the study done by Endo and colleagues, were the sources of ribosome-inactivating proteins as well as other translation inhibitors. These were considered to be responsible for the poor stability as well as the low yield of the conventional wheat germ cell – free system (Madin et al., 2000). Endo and colleagues further developed a high – yield long – lasting wheat germ cell-free system which was done by extensively washing isolated wheat germ (Madin et al., 2000; Sawasaki et al., 2007).

Till the time now, the extracts of mammalian cells along with the exception of the rabbit reticulocyte lysate have not been widely used for cell-free protein synthesis majorly because of the low yields. At RIKEN, Imataka and colleagues found that in the HeLa cell extract, the lack of functional translation initiation factors was a major limiting factor (Mikami et al., 2006b).

The development of the significantly improved mammalian cell-free systems were done by the Imataka group and this was done by supplementing with exogenous translation initiation factors (Mikami et al., 2010; Mikami et al., 2006a; Mikami et al., 2006b). Moreover, this was done by effectively taking over the rabbit reticulocyte lysate system.

Ultimately, the Ueda group reconstituted the minimal protein translational machinery from *E.coli*. This was done by completely circumventing the issues of inhibitory factors in cell extracts (Shimizu et al., 2001). The PURE system that is Protein synthesis Using Recombinant Elements was created by the Ueda group and this system consists of wholly purified components along with minimal nucleases, proteases, and other enzymatic activities which are detrimental or unrelated to protein translation.

Even though other laboratories had done many attempts in order to reconstitute the protein synthesis along with purified components (Ganoza et al., 1985; Kung et al., 1977), the PURE system was considered to be the first robust cell - free system which contained a complete set of purified translation factors and the aminoacyl-tRNA synthetases (Shimizu et al., 2001; Shimizu et al., 2005).

Keeping the same approach in mind, the laboratory reconstituted the minimal protein translational machinery from *Thermus thermophilus*. Furthermore, it demonstrated cell-free protein synthesis at temperatures up to 65°C (Zhou et al., 2012).

Swartz and colleagues pioneered chemical engineering approaches for making the *E.coli* extract, to make cell-free protein synthesis by using a protein factory and not just an analytic tool. Their main aim was to manufacture therapeutic proteins in the cell-free system at a liter or any scale (Carlson et al., 2012).

The whole processes of cell growth, extract preparation, protein synthesis and downstream purification were considered for the cost analyses as well as process engineering. The "black-box" nature of the *E.coli* extract was then acknowledged as the sets of biochemical reactions amenable to pathway engineering.

Among many key advancements, one of them was the creation of the "Cytomim" system. This system could activate central metabolism and oxidative phosphorylation in the *E.coli* extract, furthermore, allowing the use of inexpensive energy substrates, such as glutamate or glucose, and also, continuous protein synthesis and that too for more than 6 hours in the batch format (Jewett et al., 2008; Jewett and Swartz, 2004a; Jewett and Swartz, 2004b).

The Cytomim system was considered to be the basis for further improvements. These improvements resulted in the cell - free synthesis of recombinant human granulocyte-macrophage colony-stimulating factor in a 100-liter industrial stir tank (Jewett et al., 2008; Zawada et al., 2011).

2.5. STEPS TOWARD AN *IN VITRO* PROTEIN SYNTHESIS SYSTEM

In order to approach protein synthesis, the first attempt was made through tracing. It was basically for testing the animals with the radioactive amino acids that were present in the tracing process. The following step included the incorporation of the label into the proteins of different tissues. However, at that time, there were no radioactive labeled amino acids present commercially.

These radioactive *in vivo* studies were mainly concerned about the maintaining of control over the specific activity of the injected material. After that, the test tube protein synthesis systems were tested by the researchers from the animal tissues. The people from the University Of California Medical School at Berkeley were some of the people to use the tissue slices for the first time, naming some of them, Theodore Winnick, Felix Friedberg, David Greenberg, Jacklyn Melchior, and Harold Tarver.

Friedberg et al., Melchior and Tarver and Henry Borsook's team at Caltech, at that time, tried to incorporate amino acids into proteins of tissue homogenates. However, different amino acids were used by all, Tarver used sulfur-labeled cysteine and methionine, Greenberg and Winnick used carbon-labeled glycine and Borsook used carbon-labeled lysine.

The early incorporation of the amino acids in *in vivo* studies was due to the turnover reactions of the amino acids. But there was peptide bond formation in the testing of the protein synthesis in the test tube, which was considered as the biggest concern in the studies done by all during the years 1950 and 1955.

The work of Paul Zamecnik was the cutting-edge work in the field during the years 1950 and 1960. Zamecnik worked at the Massachusetts General Hospital (MGH) in Boston. An organic chemist known as Robert Loftfield, who was from Radioactivity Center at MIT, joined the staff of the Massachusetts General Hospital in 1948. Loftfield worked hard in finding a suitable method for the synthesis of 14C-alanine and glycine.

After the work of Loftfield, there were sevral researchers who started to introduce the radioactive amino acids into slices of the livers of rats. They were Warren Miller, and Ivan Frantz, Zamecnik. William Loomis showed that there was an interference of the dinitrophenol (DNP) specifically with the process of phosphorylation. It was done in the laboratory of Lipmann, who was the neighbor of Zamecnik at MGH.

The amino acid incorporation activity was stopped when there was the inclusion of DNA into the slice experiments by the Zamecnik group. According to the results, the protein synthesis was coupled with the phosphate bond energy utilization. This was however assumed for a long time by Lipmann based on peptide bond model reactions.

2.6. THE GOLDEN AGE OF TRANSLATION – THE 1960S

The year 1960 was the innovator year of messenger RNA, the resolution of the transaction translational process into partial functions and the emergence of the dissection of the ribosomes into its components. In the year between 1961 and 1965, the genetic code was solved having a velocity that is difficult to predict by anyone.

The biochemistry of photosynthesis was merged by transfer RNA, messenger RNA and the code. For some time, the biochemistry of protein synthesis became synonymous with molecular biology. It is a situation when a gap still existed between those who did the work of experimenting with the draining of bog of nucleic acid or protein biochemistry and metabolism and those who were the avant-garde of molecular biology.

The *In vitro* system remained central throughout. It was important basically for the transition of the mammalian systems to the bacterial systems. Basically, it is about the *E.coli* system of protein synthesis. *E.coli* was considered as the most vital genetic model throughout the year 1950. However, this was not the model of translation for the decade. The molecular genetics and protein synthesis research joined forces depending on the one single model organism. It was achieved in 1960.

2.7. 1970S–1990S: A BRIEF SYNOPSIS

The year 1970 was the year that witnessed the elucidation of the primary structure of the components of the translational apparatus. Moreover, at another side of the decade, it was noted that the first cellular organelle whose RNA and protein components was completely sequenced was the ribosome of *E.coli*. Maxam and Gilbert and Sangar both introduced a new sequencing method that made it a feasible task to sequence the complete ribosomal RNA.

A detailed genetic map of the components that are there in the protein biosynthesis could be constructed with the help of the emerging recombinant DNA technologies. It was first identified in the year 1962, and many years have passed for precise mapping.

It was after the subsequent isolation of the protein gene-transducing lambda phages that the knowledge regarding operons and ribosomal protein genes accumulated at a fast pace. The ribosomal protein mutants when worked with systematically provided information which was considered as another good source of information.

The details of the molecular ribosomal function also became available. The information such as the interaction of mRNA with 16S RNA during initiation and the methods through which ribosomes achieve their accuracy. During the 1970s, an other major field of investigation was the regulation of ribosome biosynthesis which started with the early findings on the genetics of RNA synthesis.

Firstly, a detailed view of the transcriptional and then of translational feedback regulation mechanisms emerged over the years. It is because of the work done by Monod and Jacob on lac operon, the leading paradigm came out to be was the transcriptional control.

In prokaryotes and eukaryotes, there was a shift of interest from the transcriptional to translational regulation. It was considered as an unprecedented turn (Severo Ochoa, The Regulation of Protein Synthesis in Reticulocytes, 1978). Nomura had initiated some major events in this area and reviewed from time to time.

There were numerous advancements and development done in the field of ribosome research. It focused mainly on the application of advanced biological techniques, biochemical, biophysical. The localization of ribosomal components (*in situ*) and reconstitution of ribosomes (*In vitro*) through affinity labeling, cross-linking, immunoelectron microscopy and scattering studies. This led to the early insights into the quaternary structure in the organelle of protein synthesis and the characteristics that are functional such as factor binding and the constitution of the peptidyltransferase center.

In the year 1980, there was an increase in the back shift of the emphasis towards the eukaryotic systems on one hand. It was in Chicago that Ira Wool, during the era of the *E.coli* pioneered mammalian ribosomal proteins. In Amsterdam, there has been how much work done on the genetic and structural work on the yeast ribosomal RNA by Rudy Planta.

After a long time, on the other hand the function of the ribosomes also included the time consuming and tedious task of three sectors as secondary tertiary and quaternary structure modeling came to fruition. The continuing efforts in this field involved various methods such as protection and modification studies, Protein–protein crosslinking, neutron scattering, protein–RNA crosslinking and Ribosome crystallization.

A new model was found using the tRNA-binding studies. On the functional side, with the help of the third tRNA binding site, it led to the conceptions of the elongation cycle. Judgement was passed by Peter More on this "The two-site model for the ribosome, which the world has accepted for a generation is dead. The existence of a third site for tRNA binding, the exit site, is now established beyond a reasonable doubt. This is unquestionably the most significant advance in our understanding of the ribosomal events of protein synthesis in many years".

The structural analysis of the ribosomes to the atomic resolution was seen in the year 1990s. It was done with many people's efforts. The solution of the phasing issue and the ribosomes and ribosomal subunits contained suitable crystals that were mainly from the sources of thermophilic and halophilic led to the proliferation of X-ray crystallographic studies. In the year 1980s, the initiative was taken by Wittmann and Ada Yonath in Berlin and the group at Pushchino, that laid the ground with ribosome crystallization.

The continuous efforts of around 20 years have brought fruitful results for the large ribosomal subunit. The atomic resolution for the large ribosomal subunit has been achieved by *Haloarcula marismortui* and *Deinococcus radiodurans*. The atomic resolution for the small ribosomal subunits has been achieved by *Thermus thermophilus* and near-atomic resolution for the 70S-tRNA–mRNA complex.

The era for the ribosome at atomic resolution came around in the year 2000 and Nobel prize awarded to Ada Yonath, Venkatraman Ramakrishnan and Thomas Steitz in the year 2009. The development of the reconstruction of the cryoelectron microscopic image parallelly has refined particularly the specific functional states of the overall 3D shape of ribosomal particles.

2.8. PROTEIN SYNTHESIS AND TRANSLATIONAL CONTROL: A HISTORICAL PERSPECTIVE

The regulation of protein synthesis and the protein synthesis itself is central to all the forms of life and is applied in varied arenas as agriculture, biotechnology, and medicine. Since the mid of the 20th century, translation and translational activities have seen a big increase in its investigation intensity. There is an increase in the advances in molecular and cell biology.

In a cell, the largest amount of the macromolecules is the proteins that are considered as the important components of the extracellular milieu. Protein fulfills various roles such as transport, enzymatic, structural, and other roles as well in the entire organism. The synthesis of the proteins requires extensive biological machinery and there is a need for delicate and sophisticated regulation to accomplish this. The synthesis is done through the translation of genetic information which is encoded in the messenger RNA (mRNA)(Hershey et al. 2018).

The modulation of the protein synthesis is done quantitatively and time and space, with the help of a network of stimuli, responses, and interactions collectively referred to as translational control. As discussed before in terms of genetics, cell biology and bioenergetics, translation and translational control include a large proportion of the resources of cells and organisms (Mathews et al. 2007).

2.8.1. Translation Timeline

The early 1950s noted that the research of the protein synthesis mostly addressed the findings that were descriptive in nature and there were mostly questions that arise physiologically (Daly and Mirsky 1952). The template theories assisted the theories of protein synthesis through enzyme assembly and peptide intermediates (Campbell and Work, 1953). At that time, there was no translation system known (Palade, 1955). The studies of protein synthesis were performed in the tissue slices and in the whole animals in this era.

The decade of 1950s noted many crucial advances in technical terms (such as radiolabelled amino acids, sucrose gradient centrifugation, and inhibitors), the development of cell system and some fundamental discoveries (such as of transfer RNA [tRNA] and ribosomes) which were the transformations for the field. The decade after the 1950s was dominated by many other fields such as themes, concepts, and discoveries.

The seminal elucidation of the genetic code was done in the year 1960 and along with this, there were some foundational discoveries in the area of polysomes and mRNA which continued to be made in the same year. 1975 showed the beginning of the mechanic studies and there was a rapid growth in the field with the most of the characterization of the components of the eukaryotic translation system. The subsequent years of 1980 so the rapid increase in the regulatory and mechanistic themes. In the 1990s, the expansion and appreciation of regulatory pathways were seen. The current activity of the 21st century is derived from these things which have contributed a lot in its first decade. Although there have been many types of research being done and still being in the process which tries to explain the pathophysiological responses and physiological responses of the translation system.

The research also focuses on the treatment and to develop therapeutic of the genetic and acquired infections or diseases. There has been chronological and theme article record of the principal discoveries which are distinguished by color coding. The research shows that this field does not show any progress in a systematic and orderly level as in other fields.

The development of the theories and discoveries has been very slow and some even lay ideal for many years. A good example to support this is the role of eukaryotic initiation factors if that has similar cases like this. Some other discoveries such as poly (A) which is considered as the tale of eukaryotic mRNA has far-reaching effect in several areas of mRNA characterization, metabolism, translation, and isolation.

It rapidly extended in advancement too many other areas as well. The greater depth studies and the advancement in the field were benefited from these studies. These depth studies were being enabled by new techniques such as blotting, toe printing, ribosome profiling, electrophoresis and some technological development in the same field such as reverse genetics, cloning and x-ray crystallography.

This led the field to research more about it which letter on provided the mechanic understanding and the fresh insights about the detailed clarity. The structural investigations regarding ribosomes and the interactions with the translation system components are some of the examples of the above (Jobe et al., 2018).

However, sometimes these investigations and studies in the same field resulted in the observations that were not expected which included unorthodox initiation mechanism (Meyer et al. 2015; Zhou et al. 2015; Kwan and Thompson 2018; Zu et al. 2018) and novel concepts such as ribosomal heterogeneity (Sauert et al. 2015; Genuth and Barna 2018). The scope of the field was expanded due to the ribosomal concentration model (Mills and Green 2017; Khajuria et al. 2018).

2.9. DEVELOPMENT OF THE PROTEIN SYNTHESIS FIELD

The field of protein synthesis started its biochemical investigation at the time when the concepts that can now be written in an axiom form were uncertain. The first protein sequence was about to be established and published which viewed proteins as the unique linear arrays of about residence of 20 amino acids (the insulin B chain, Sangar and Tuppy, 1951).

The establishment of the protein was also viewed as the mechanism of its synthesis which also involved the reversal of proteolysis or phosphorylated intermediates that were entertained at that time (Zamecnic, 1969).

In late 1940, there was a revolution in many areas of biomedical science and especially the radioactive isotopes. The radiolabeled amino acids traces came into use in the year around 1950. The simple labeled compounds were synthesized by the researchers as a four-step in their experiments. They synthesized it until they become commercially available in the latter decade. The examples of simple- labelled compounds are as formaldehyde or sodium cyanide (Borsook et al., 1952).

After the invention of this profound technical advancement, there was speedy investigation regarding the subject that led biochemistry to stand ahead of genetics. However, biochemistry stood ahead of genetics until the time cloning was invented. The east system was also exploited systematically which started its journey in the 1980s.

There was a production of a Cell-free preparation that was done from rat liver. The it incorporated the amino acid into protein and hence it showed that ATP and GTP are required as an energy form (Zamecnik and Keller, 1954; Keller and Zamecnik, 1956).

There were some stages in the translation system that was refined and resolved into subfractions. It also included a microsomal fraction containing ribosomes which were attached to intracellular membrane fragments (Zamecnik, 1960). The process of protein synthesis in bacterial cells is very rapid and there is a pulse -chase experiment that showed the ribosomes are the site of protein synthesis. It took only 5 to 10 seconds in the assembly of a protein chain on a ribosome (Mcquillen et al., 1959).

However, the central role played by RNA in the flow of genetic information to protein was accomplished in advance of its understanding. This was before the time of the visualization of polysomes and when before the first RNA sequence was completed (Holley et al, 1965).

Among all the concepts, one hypothesis says that "one gene-one ribosomes-one protein" according to which each ribosome is dedicated to the synthesis of a single protein. The polysomes and the functions of the polysomes were observed and in the early years of the 1960s.

It was also appreciated in the light of the messenger hypothesis and the "tape mechanism" of translation (Marks et al. 1962; Warner et al. 1962, 1963; Arlinghaus and Schweet 1963; Gierer 1963; Goodman and Rich 1963; Nakamoto et al. 1963; Noll and Wettstein 1963; Wettstein et al. 1963). This developmental field made numerous contributions as the technical advancement in electron microscopy (EM) and high-speed centrifugation.

The establishment of aminoacyl-tRNA and its role was done in the late 1950s. The detection of the intermediate and activated state of amino acids was done firstly (Hultin and Beskow 1956) and then the characterization of the activated state of amino acids was done (Hoagland et al. 1958, 1959). It was then recognized and predicted on theoretical grounds as the physical manifestation of the adapter RNA (Crick 1958).

After the realization of its functions, the name of the transfer RNA was replaced by Soluble RNA (sRNA). The decoding of the template of the RNA component is due to the chemical modification of the amino acid moiety of cysteine-charged $tRNA^{Cys}$ (to alanine) (Chapeville et al. 1962).

The aminoacyl-tRNA synthetases is responsible for the fidelity of information transfer from nucleic acid to protein. It was later found that valine-specific enzyme from *Escherichia coli, was the* first macromolecular component of the protein synthetic apparatus that was characterized (Berg and Ofengand 1958). Thereafter, the additional synthetases were followed.

In the year 1960s and 1970s, there was a purification of numerous enzymes that were catalyzing and facilitating several steps in protein synthesis. Before these proteins were fully purified and the full authentication of its functions, these were provisionally called "factors".

Still, there are some factors that have resulted in the activity that remained debatable and obscure for almost half a century. For example, EF-P and its homolog eIF5A (Kang and Hershey 1994; Aoki et al. 1997; Doerfel et al. 2013; Ude et al. 2013). On one hand, some are still emerging and some having specialized functions such as trans-acting factors (ITAFs) and internal ribosome entry site (IRES), are still discovered (King et al. 2010; Lee et al. 2017).

All the cells containing the gene expression has changed thinking due to the revolutionized concept of messenger RNA. The main key role has been played by genetics, bacteriophage biology and biochemistry in the genesis and confirmation of the messenger hypothesis (Cobb, 2015). The hypothesis was done by Jacob and Monod (1961) who studied the existence of an unstable intermediate between the DNA of the gene and the ribosome. It is to be noted here that it could be related to the RNA produced in phage T2-infected cells (Volkin and Astrachan 1956; Nomura et al. 1960).

Considering this view, it can be said that the ribosome and other components of the protein synthesis machinery contain a relatively stable decoding as well as synthetic apparatus which is programmed by unstable mRNA. Soon, this was confirmed in bacteria as well as bacterial cell – free systems (Brenner et al. 1961; Gros et al. 1961).

There was a discovery that the poly (U) can provide directions to the synthesis of polyphenylalanine *in vitro* (Nirenberg and Matthaei, 1961). This discovery was a transformative event while spearheading the elucidation of the genetic code by the period of the mid – the 1960s. There was a wobble hypothesis and this hypothesis rationalizes the features of the redundancy of the code and its decoding by tRNAs. This wobble hypothesis was published by Crick (1966). Considering the higher cells, it was further recognized that there was the existence of a class of rapidly labeled RNA was heterogeneous in size and having distinct chromatographic properties.

It has an essential feature and that is informational intermediaries which were confirmed and messenger – dependent eukaryotic cell – free translation systems that appeared at the end of the decade (Laycock and Hunt 1969; Lockard and Lingrel 1969; Mathews and Korner 1970).

Many mechanisms were built on these foundations which further explain some of the important aspects of translation in both prokaryotes as well as eukaryotes. This emerged in the period of 1970s. There a question arose which was of initiation site selection and this largely accounted for in prokaryotes by the Shine – Dalgarno sequence base – pairing along with 16S ribosomal RNA (Shine and Dalgarno 1975).

For the problem in eukaryotes, a solution came later in the period of the 1980s along with the recognition of cap dependent scanning and the Kozak consensus sequence (Kozak and Shatkin, 1978). Further, this was followed by the identification of IRES-dependent mechanisms (Jang et al. 1988; Pelletier and Sonenberg 1988).The optimal codon hypothesis further explained a bias in the usage of synonymous codons. This can influence the production levels of individual genes (Ikemura, 1981) and the signal hypothesis which considered the protein transport into the ER that is the endoplasmic reticulum.

The facilitation of the investigations of mRNA translation and the metabolism was done by the discovery in the early period of the 1970s of the terminal hallmarks which was found on most eukaryotic mRNAs 5′ caps (Adams and Cory, 1975; Both et al. 1975; Furuichi and Miura 1975; Furuichi et al. 1975a,b; Perry and Kelley 1975; Wei and Moss 1975) and 3′ poly(A) tails (Darnell et al. 1971a,b; Edmonds et al. 1971; Lee et al. 1971).

By the ending period of the 1980s, the identification and purification of most of the components of the translation system and the reconstitution of their activities *in vitro* further resulted in the pathway of protein synthesis and this had been defined. Moreover, this was well understood in the outline (Hershey et al. 2018).

There were subsequent detailed analyses that resulted in an in-depth understanding of a number of mechanisms of initiation, elongation, and termination in bacteria (Rodnina, 2018). Considering the eukaryotic cells, this was done along with indispensable contributions from yeast genetics as well as biochemistry (Dever et al. 2018; Hellen 2018; Merrick and Pavitt 2018). This has given rise to many questions related to regulation that need to be addressed at levels of sophistication which is ever-increasing.

2.10. ORIGINS OF TRANSLATIONAL CONTROL

There was a natural flow of the idea from the concept of messenger regarding regulation at the transcriptional level. According to Jacob and Monod (1961), "the synthesis of individual proteins may be provoked or suppressed within a cell, under the influence of specific external agents, and … the relative rates at which different proteins are synthesized may be profoundly altered, depending on external conditions."

Such regulation was recognized as "is essential to the survival of the cell,". However, the notion that favors the exertion to the translational level was not basically what was being on focus in the bacterial field. It has to be noted that the seeds of the concept that the thing behind the regulation of the efficiency of protein synthesis is the gene expression emerged early. Some of them came from bacterial systems.

The first observation of the regulation was made in the rabbit reticulocytes at the level of gene expression (Borsook et al, 1952; Kruh and Borsook,

1956). The term "translational control" was used in reference to the proteins of differential expression with the RNA genome of MS2 phage in an *E.coli* cell-free translation system in the year 1963 (Ohtaka and Spiegelman, 1963).

However, the concept spread very quickly in different areas of research work. It was extended to such an extent that the presence of silent mRNA translationally which was activated on fertilization of sea urchin eggs were referred to as "classical conclusion" (Humphreys, 1971). Thereafter, at the early studies of phases, there was more focus on eukaryotic systems.

There is a rapid response of the external stimuli without invoking nuclear pathways for mRNA synthesis, transport, and processing. This is due to the virtue of translation as a site of regulation. On the other hand, most of cases recognized were the ones where there was no evidence. It could be simply understood as the establishment that transcription and other nuclear events were not responsible.

2.11. CONCLUSION

It is anticipated that the coming years will increase the utility of the cell-free protein synthesis (CFPS) systems and it will only expand. The reason being the production of the high output of the protein production, high-level of protein production, cost-effective and high throughput. One of the main challenges is that there is no single platform to reliably synthesize any biologically active protein.

This includes the lack of cost-effective and scalable eukaryotic cell-free protein synthesis platform. The humans cannot carry out glycosylation patterns. These challenges when addressed are limited not only by the technicalities in the field but by the increasing number of applications that help in cell-free protein synthesis.

Cell-free protein synthesis (CFPS) systems can be modified and controlled. The emergence of the Cell-free system on the industrial scale, the applications of cell-free protein synthesis have now come of age. However, the field is still progressing and is only beginning to reach its full potential.

REFERENCES

1. Carlson, E., Gan, R., Hodgman, C. and Jewett, M., 2012. Cell-free protein synthesis: Applications come of age. *Biotechnology Advances*, [online] 30(5), pp.1185-1194. Available at: <https://www.ncbi.nlm.nih.gov/pmc/articles/PMC4038126/> [Accessed 6 May 2020].
2. Khambhati, K., Bhattacharjee, G., Gohil, N., Braddick, D., Kulkarni, V. and Singh, V., 2019. Exploring the Potential of Cell-Free Protein Synthesis for Extending the Abilities of Biological Systems. *Frontiers in Bioengineering and Biotechnology*, [online] 7. Available at: <https://www.frontiersin.org/articles/10.3389/fbioe.2019.00248/full> [Accessed 6 May 2020].
3. Mediatheque.lindau-nobel.org. n.d. *Topic Cluster - A Brief History of Protein Biosynthesis and Ribosome Research*. [online] Available at: <https://www.mediatheque.lindau-nobel.org/topic-clusters/a-brief-history-of-protein-biosynthesis-and-ribosome-research#page=2> [Accessed 6 May 2020].
4. Tahmasebi, S., Sonenberg, N., Hershey, J. and Mathews, M., 2018. Protein Synthesis and Translational Control: A Historical Perspective. *Cold Spring Harbor Perspectives in Biology*, [online] 11(9), p.a035584. Available at: <https://cshperspectives.cshlp.org/content/11/9/a035584.full> [Accessed 6 May 2020].
5. Anderson et al., 1983
6. Aoki et al. 1997
7. Arlinghaus and Schweet, 1963
8. Bank and Marks, 1966
9. Berg and Ofengand, 1958
10. Borsook et al., 1952
11. Brenner et al. 1961
12. Carlson et al., 2012
13. Chen and Zubay, 1983
14. Cobb, 2015
15. Craig et al., 1992
16. Crick, 1970
17. Doerfel et al. 2013
18. Erickson and Blobel, 1983

19. Gale and Folkes, 1954
20. Ganoza et al., 1985
21. Gesteland, 1966
22. Gierer, 1963
23. Goodman and Rich, 1963
24. Gros et al. 1961
25. Hershey et al. 2018
26. Humphreys, 1971
27. Jackson and Hunt, 1983
28. Jacob and Monod, 1961
29. Jewett and Swartz, 2004b
30. Jewett et al., 2008
31. Kang and Hershey, 1994
32. Keller and Littlefield, 1957
33. Keller and Zamecnik, 1956
34. Kim et al., 2006
35. King et al. 2010
36. Krieg and Melton, 1987
37. Kruh and Borsook 1956
38. Kung et al., 1977
39. Lamborg and Zamecnik, 1960
40. Lee et al. 2017
41. Liu et al., 2005
42. Madin et al., 2000
43. Marcus and Feeley, 1966
44. Marks et al. 1962
45. Matthaei and Nirenberg, 1961
46. Michel-Reydellet et al., 2004
47. Mikami et al., 2010
48. Nakamoto et al. 1963
49. Nevin and Pratt, 1991
50. Nirenberg and Matthaei, 1961
51. Nomura et al. 1960

52. Ohtaka and Spiegelman 1963
53. Pelham and Jackson, 1976
54. Pratt, 1984
55. Roberts and Paterson, 1973
56. Ryabova et al., 1998
57. Sangar and Tuppy, 1951
58. Sawasaki et al., 2007
59. Schweet et al., 1958
60. Schweet et al., 1958
61. Shimizu et al., 2005
62. Shirokov et al., 2007
63. Spirin and Schwartz, 2008
64. Spirin et al., 1988
65. Strasser, 2006
66. Ude et al. 2013
67. Volkin and Astrachan, 1956
68. Warner et al. 1962, 1963
69. Wettstein et al. 1963
70. Yang et al., 1980
71. Zamecnik and Keller, 1954
72. Zamecnik et al., 1948
73. Zamecnik, 1969
74. Zawada and Swartz, 2006
75. Zawada et al., 2011
76. Zhou et al., 2012
77. Zubay, 1973

3

t-RNA and Aminoacyl-tRNA Synthetases

CONTENTS

This chapter is about the structure of tRNA, its role. This chapter gives details about the role of tRNA. For explaining it in detail, the multiple biological functions of tRNA are being explained in detail.

This contains, tRNA cycle in protein synthesis, tRNA and regulation of enzyme synthesis, likewise there are many other procedures for explaining the biological function of tRNA.

With this, likewise, the aminoacyl-tRNA synthetases have some reaction steps that are also being explained in this chapter in detail. This chapter explains the role of tRNA and aminoacyl-tRNA in the process of encoding the genetic code of messenger RNA.

3.1. INTRODUCTION

Transfer RNA (tRNA) is said to serve as the nucleic acid decoding device that helps in reading the triplet genetic code of messenger RNA (mRNA). They also help in the insertion of codon specific amino acids in a growing protein chain during the process of translation which occurs in the ribosome.

tRNA helps in reading the particular triplet codon that is present in the mRNA with the help of an anticodon loop, which has a triplet anticodon residue that is found to base pair with the codon. It is observed that each tRNA is being charged with the particular amino acid at its 3' end.

Even though there are 61 codons available in the universal genetic code that is specifying the amino acids. It is believed that most of the organisms possess fewer than 45 different types of tRNAs. Some of the tRNAs can be also used to read different codons.

This type of reading is possible because of the flexibility that is present in the base pairing of the third (3') residue of the mRNA codon and the first (5') residue of the anticodon. With the help of these "wobble" base pairs, which allows the non-Watson-Crick Hydrogen bonding and therefore, it allows a single tRNA to read the multiple codons.

The molecule which is present in the left is $tRNA^{phe}$, this is considered to carry the amino acid phenylalanine that is attached to its 3' end when it is appropriately charged by the tRNA synthetase enzyme.

In the past few years, the Research in the field of transfer RNA (tRNA) is believed to experience a vast revolutionary change. Even though, there has been a stable accumulation in the chemical and biological information that is regarding this molecule for about 20 years.

Until the year 1973, there was no firm information that was available in the three - dimensional structure of the molecule. In early 1973, the polynucleotide chain of yeast $tRNA^{phe}$ has been traced in a 4--X ray diffraction analysis.

Structural work is noticed to progress rapidly, from then to the point where the atomic coordinates is now available. As these coordinates are derived from 2.5.-X ray diffraction analyses from the two different crystal forms of the same molecule.

Knowledge about the detailed three-dimensional structure of the molecule that makes a change that occur separately in the type of research that can be also carried out. The main aim behind describing some of the details in the manner to obtain knowledge about the three -dimensional structure of tRNA species. they also want to discuss the extent to which it can be explained and can make various aspects understandable that are involved in chemistry and solution behavior of the same and tRNA species.

The major biological function of the tRNA is related to the role of the tRNA in protein synthesis. The existence of a molecule like tRNA is a sense that is made necessary by the fact that nature encodes the information about the genetics in the nucleotide sequence in the nucleic acid.

This helps in expressing the biological information in the ordered sequence in the polypeptide structure of amino acids. Transfer RNA is said to have a fundamental biological role which is being used in acting as an interface between the polypeptide and polynucleotides.

tRNA works by interacting with the messenger RNA in the ribosome at one end. While at the other end it is containing the growing peptide chain. tRNA is often seen to be involved in a large number of biological processes.

Aminoacyl-tRNAs are the substrates that are used for the translation process and they are considered as an essential for determining how the genetic code is being interpreted as an amino acid. The main function of aminoacyl-tRNA synthesis is to match the amino acids precisely with the tRNAs who are containing the corresponding anticodon.

This function of aminoacyl-tRNA is achieved primarily by the direct attachment of amino acids with the corresponding tRNA with the help of aminoacyl-tRNA synthetase. There are other essentials those who are beneficial in several cases, these are intrinsic proofreading and extrinsic editing. According to some recent studies on aminoacyl-tRNA synthesis, this is mainly prompted by the advent of whole- genome sequencing and by the availability of a structural data body. These studies have led towards a more detailed and expanded picture of how the aminoacyl-tRNAs are being synthesized.

Basically, the aminoacyl-tRNA synthetases (AARSs) is said to constitute a family of enzymes that help in catalyzing the covalent attachment of amino acids to their cognate tRNA during the process of translation. In the year 1951, the tRNA synthase identified that ATP is needed for the incorporation of amino acid *in vitro* into a polypeptide.

Late, in the year 1955, aminoacyl adenylate was identified. In the same, Francis Crick even introduced a hypothesis related to an adaptor. Francis proposed that the amino acid is often carried to the template by an adaptor molecule that fits into RNA. Therefore, in the simplest form, it can be said that one would require the twenty adaptors, which is one for each amino acid.

This hypothesis of adaptor got to coincide with the tRNA molecule which can be formally referred to as sRNA. Hoagland et al. hypothesized this theory in the year 1957. The complete sequence of tRNA was first published in the year 1965. Nearly a decade later, the structure of $tRNA^{phe}$ was being determined in the ear 1974.

In the first step, there is specific recognition of AASR that cognate amino acid in the presence of ATP and they also form a complex related to the enzyme-amino acid-AMP complex by simply releasing the pyrophosphate.

In the next step, for the AARs the specific tRNA gets bind to the enzyme amino acid AMP complex. With this, they form a covalent bond with the amino acid by releasing the AMP to form an amino acid enzyme tRNA complex. There are at least 20 different AARSs that are present one for each of the amino acids, and therefore there is a minimum of 20 different tRNAs which available currently.

3.2. STRUCTURE OF TRNA

The main role of tRNA is to have two-fold, firstly, bind and accept the correct amino acid from the suitable tRNA synthetase and secondly, to recognize and then bind to the complex of ribosome-mRNA to deliver the amino acid that is suitable for growing the polypeptide chain.

As it plays a central role in gene expression, therefore, tRNA is one of the best molecules to be understood. tRNA is considered as the most carefully studied biological macromolecules. tRNA contains approximately

75 to 90 nucleotides in length and with the four normally occurring RNA bases it also contains variant bases.

Most of the characteristics are the result of the products that are achieved by alkylating the adenine, guanine and cytosine amino groups. Alkylation takes place of the ring N of the bases with glycosylation and saturation of C=C bonds.

As these modifications are often restricting the type of base pairing, even these base pairing is considered to assume and have interesting consequences for the tertiary structure of tRNA. Depending on the base-base complementarity, the secondary structure of the tRNA can be even drawn as the structure of cloverleaf of 4 stems which is almost familiar. Here each of these 4 stems is made of four to seven Watson-Crick type base pairs.

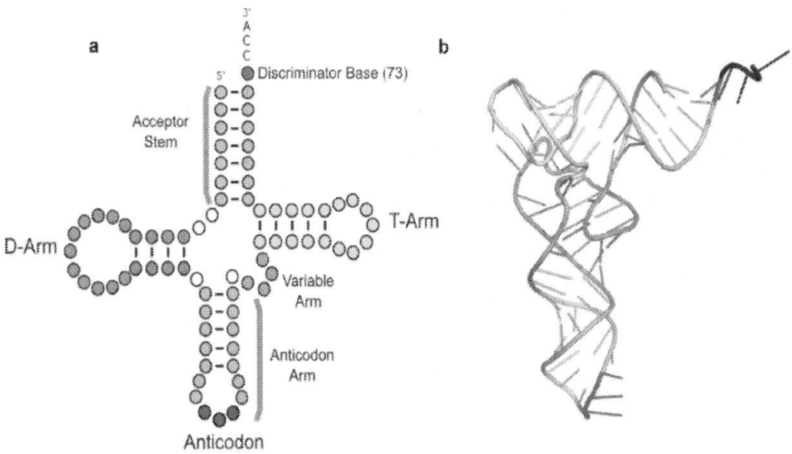

Figure 3.1. An image representing tRNA cloverleaf structure

Source: Image by ResearchGate

It is considered that the five regions of the tRNA are not base- paired, the CCA acceptor stem, the anti-codon loop and the D-loop which is named due to presence of dihydrouridine, which gets interacted with the mRNA, extra arm and the TYC loop which is named so because of the presence of the pseudo-uridine base.

Three- dimensional structure of the tRNA molecule is not a cloverleaf. As the molecule folds back on itself and further forming two segments of the double helix and in which most of them are unpaired bases in the T and D loops which form pairs with each other.

In this molecule, one of the two segments forms a double helix by the process of stacking of base pairs regions of the T and acceptor stems. The other segment is formed by stacking the D-arm base- paired region on the base- paired part present on the anticodon loop.

Therefore, this tertiary structure of the molecule is basically in the L-shaped and is being held by the base pairs that are between the bases in the unpaired region present on the stems and by the unique stacking interaction.

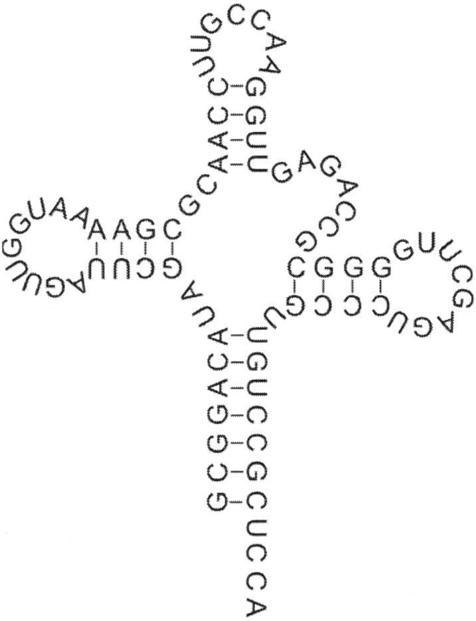

Figure 3.2.Structure of tRNA

Source: Image by Free SVG

3.3. THE MULTIPLE BIOLOGICAL FUNCTIONS OF TRNA

Even though, in the protein synthesis, the role of tRNA is usually emphasized. As it is very important to recognize that this tRNA molecule is involved with many of the other biological functions.

3.3.1. tRNA Cycle in Protein Synthesis

During the process of protein synthesis, tRNA interacts with the large number of different proteins that are considered to play an important role in its biological functions. The entire tRNA molecule ends in a common sequence that is CCA, which is added by the nucleotidyltransferases enzyme to the 3' -end of the molecule.

There is an important step in the process of protein synthesis which is specific aminoacylation. This process is carried out by the 20 different tRNA-aminoacylating enzymes or aminoacyl tRNA synthetases. These enzymes are recognizing only some specific set of iso-acceptors tRNAs which are working as a substrate.

It is believed that they require ATP for the initial activation of the amino acids before it gets transferred onto tRNA. Although the amino acids are being added to the 3' -terminal adenosine, recently it is being found that some of these enzymes like aminoacylate are found on 2' hydroxyl and some are found on the 3' hydroxyl groups.

Aminoacyl tRNA (aa-tRNA) is then carried into the ribosome complex with the help of transfer factor EF-TU which is present in prokaryotes and EFI is present in the eukaryotes. It is important to note that the initiator $tRNA^{Met}_{f}$ is considered having its factor for the ribosomal insertion.

Inside the ribosome, tRNA often gets interacted with the number of ribosomal proteins including the peptidyl transferase, before it gets finally released from the ribosome. After that, the amino acids are transferred to the growing polypeptide chain of an adjacent tRNA.

Even though, there is a fair amount that is known about the various aspects of tRNA biosynthesis and function which occur during the protein synthesis. Therefore, there is nothing that can be known similar manner in which a tRNA molecule is degraded.

3.3.2. tRNA and the Regulation of Enzyme Synthesis

The important feature of aa-tRNA is the role which is played for regulating the transcription of messenger RNA (mRNA) for the association of enzymes with the process of biosynthesis of its amino acid. This was considered to be first discovered in the operon for the histidine biosynthesis.

Recently, the regulatory role of tRNA is being reviewed, even though it is considered that most of the regulatory studies are being carried out on the prokaryotic system. According to some recent demonstration, aa-tRNA in the mammalian system helps in regulating amino acid biosynthesis.

3.3.3. Aminoacyl-tRNA Transferases

Aminoacyl-tRNA transferases are considered as a group of enzymes that help in catalyzing the transfer of amino acid from aa-tRNA to a specific acceptor molecule that occur without the participation of the ribosomes or of other kinds of nucleic acid.

The acceptor molecules can be divided into further three classes:

- When the acceptor acts as an N-acetyl muramyl peptide then there is an intermediate synthesis of interpeptide bridges that are present in bacterial cell walls.
- When the acceptor can act as an intact protein in which the amino acid is being added to the N-terminus of the protein.
- The acceptor can act as phosphatidylglycerol molecule in which the enzyme catalyses the formation of the aminoacyl esters of phosphatidylglycerol which are considered as a component of the cell membrane.

These are considered as an important link in the process of biosynthesis of the cell wall, for this purpose the specialized tRNA is being used as tRNA transferases.

3.3.4. tRNA Participation in Polynucleotide Synthesis

Reverse transcriptase is an enzyme that is found in the oncogenic viruses that are being used for making the DNA copy of the viral RNA. In this process, it has been found that there are particular species of tRNA that are being used as a primer.

$tRNA^{Trp}$ is being used by the avian myeloblastosis reverse transcriptase. The murine leukemia virus enzyme uses $tRNA^{pro}$ as a primer. According to the recent studies it being shown that, reverse transcriptase contains a strong affinity towards the tRNA primer.

There is an interesting finding that bears some of the relationships that tell the fact that there are many plant viral RNAs that possess a “tRNA-like” structure present at the 3’ -end of the RNA. There are many plant viral RNAs as well as there are many animals’ viral RNA that are found to act as a substrate for the aminoacylation by the aa-tRNA synthetases.

Haenni and co-workers in the year 1975 through their work suggested that the bacterial viral RNAs also possess some of the features of "tRNA-like" structures, not on the 3' -end. It is considered that one of the proteins was to bind specifically to the aa-tRNA. The transfer factor EF-Tu is considered as a component of the enzyme, Qβ replicase which is seen to involve in the replication of the bacterial viral RNA.

These "tRNA-like" structures are found to be present in many of the plant and bacterial viral RNAs, which play an important role in the specific recognition of these RNAs by the corresponding RNA replicases. It is considered as an interesting possibility that is needed to be explored further.

3.3.5. tRNA as an Enzyme Inhibitor

tRNA is considered as a potent inhibitor of *E.coli* endonuclease I. The work done by Goebel and Helinski (1970)suggested that tRNA alters the mode of action of endonuclease I from the double strand scission of DNA towards the nicking activity. In Drosophila, it has been found that there are specific iso-acceptor species of $tRNA^{Tyr}$ that is found to act as an inhibitor towards the enzyme tryptophan pyrrolase (Jacobson, K. B. 1971).

This is found to be involved in the conversation of tryptophan as an intermediate that is present in brown pigment synthesis. In this case, there is an uncharged tRNA that acts in the regulatory capacity. This is made possible by directly interfering with the individual enzymatic activity.

Though, there is an alternative that is used for the explanation which has been proposed recently (Mischke, D., Kloetzel, P., Schwochan, M., 1975).

3.3.6. tRNA Changes in Cells

There is a large literature that is dealing with the changes that have been observed in the cell content of tRNAs. Two review articles help in summarizing the variety of results that are dealing with the change of tRNA that is occurring in the embryogenesis during the various stages of development.It is not even clear whether these will reflect an expression of the role of tRNA which will help in regulating the systems. These systems can be either involved in the modification or regulation of other functions as well.Additionally, there is a substantial literature that is reviewed in Cancer Research Dealing with the changes in the tRNA during the process of oncogenesis. The entire volume is being devoted to this particular subject (Cancer Res., 1973).

There is a question that arises regarding the use of tRNA in a large variety of biological functions? This class of molecule is involved in the biochemistry of the living organisms from the evolutionary process and it can also find that nature is opportunistic in using such molecules for different purposes.

tRNA interacts with the protein molecule in a highly specific manner which is among a large number of biological functions. As the nature of these interactions is often unknown. But the interaction is said to involve the recognition of the tRNA molecule as distinct from the other species of RNA. This interaction takes place by the simple three-dimensional folding of the molecule and by the detection of a specific nucleotide sequence or nucleotides in tRNA by a large number of proteins.

This will be helpful for understanding the extent to which the molecular structure will serve for understanding about the detailed manner in which tRNA will interact with the variety of protein while carrying out the larger number of different biological functions.

3.4. STRUCTURE AND CLASSIFICATION OF AARSS

The process of aminoacylation is considered to occur at three locations in humans. Firstly, the process of aminoacylation occurs in the nucleus, to charge the nuclear -encoded tRNA with the proper amino acid. After this, the charged tRNA-amino acid complex enters the cytoplasm to continue the process of translation of the protein.

Secondly, the aminoacylation process occurs in the cytoplasm and for the process of protein synthesis, it enters the ribosome. The process of protein synthesis also occurs in the mitochondria, which tells that the AARS activity is also necessary for the mitochondria.

In order to facilitate the process of protein synthesis, the nuclear- encoded AARS is then transported to the mitochondria for the process of translation, which basically utilizes the specific localization signal of AARS proteins.

Further, based on these locations, the AARS is classified into two more types: mitochondrial AARSs and cytoplasmic AARSs. In the mitochondrial AARSs genes are being marked with the same code as in case of cytoplasmic AARSs with the suffix "2".

There are about 36 AARS that are being identified so far that charge all the tRNAs in the humans. Among these AARSs, there are 16 AARSs functions are occurring entirely in the cytoplasm and 17 of them are said to function exclusively in the mitochondria; whereas, 3 AARSs are found to function in both the organelles.The charging of tRNA for the Glu and Pro is carried out by a single enzyme which is present in the cytoplasm. Therefore, it is believed that both the organelles have the AARS that are used for charging all the 20 amino acids.

Further, based on the high -resolution crystal structure of AARSs, there is a further classification of these 20 AARSs classes: Class I and Class II (which is shown in Table 3.1.) that are considered to be constant throughout the evolution.

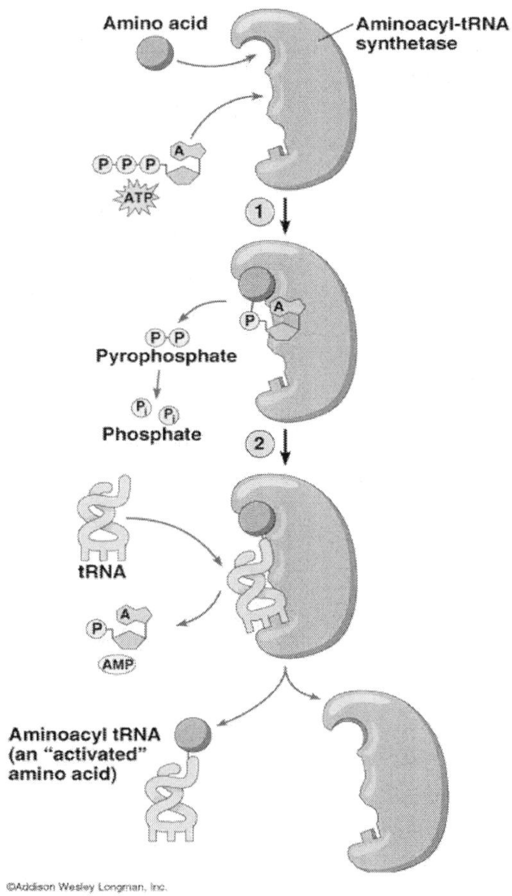

Figure 3.3. An image representing the procedure of AARS

Source: Image by Flickr

Table 3.1. Classification of Aminoacyl tRNA Synthetases

Class	Sub-Class	AARS (Form)	Organism	PDB ID
I	a	Isoleucyl tRNA synthetase (α) Methionyl tRNA synthetase (α2) Leucyl tRNA synthetase (α) Valyl tRNA synthetase (α)	*Thermus thermophilus* *Thermus thermophilus* *Thermus thermophilus* *Thermus thermophilus*	1ILE 1A8H 1H3N 1GAX
	b	Arginyl tRNA synthetase (α)	*Thermus thermophilus*	1IQ0
	c	Cysteinyl tRNA synthetase (α) Glutaminyl tRNA synthetase (α) Glutamyl tRNA synthetase (α)	*Escherichia coli* *Saccharomyces cerevisiae* *Thermus thermophilus*	1LI5 4H3S 1J09
	d	Lysyl tRNA synthetase I (α)	*Pyrococcushorikoshii*	1IRX
	e	Tyrosyl tRNA synthetase (α2) Tryptophanyl tRNA synthetase (α2)	*Thermus thermophilus* *Homo sapiens*	1H3F 2AKE
II	a	Glycyl tRNA synthetase (α2) Prolyl tRNA synthetase (α2) Histidyl tRNA synthetase (α2) Serinyl tRNA synthetase (α2) Threonyl tRNA synthetase (α2)	*Thermus thermophilus* *Plasmodium falciparum* *Escherichia coli* *Methanosarcina barkeri* *Escherichia coli*	1ATI 4TWA 1HTT 2CJB 1QF6
	b	Alanyl tRNA synthetase (α4) Glycyl tRNA synthetase (α2β2) Phenylalanyl tRNA synthetase (α2β2) Pyrolysyl tRNA synthetase (α2)	*Archaeoglobus fulgidus* *Homo sapiens* *Thermus thermophilus* *Methanosarcinamazei*	3WQY 2ZT5 1B70 2Q7H
	c	Asparaginyl tRNAsynthetase(α2) Aspartyl tRNA synthetase (α2) Lysyl tRNA synthetase II (α2)	*Brugiamalayi* *Saccharomyces cerevisiae* *Homo sapiens*	2XGT 1ASY 4YCU

Source: V. Rajendran et al (2018)

The class I enzyme family is considered to share the conserved HIGH and KMSKS motifs that they are in the active site of the Rossmann Fold. This Rossmann Fold is made up of alternative α-helices and β-sheets that are involved in the ATP binding at the time of extended conformation.

The HIGH motif is found in the loop that is joining the first β-stand to the subsequent of the α-helix (which are commonly stranded A and helix B). the KMSKS motifs are located in the loop which is immediately following the fifth β-stand.

The class I synthetases are usually found to be monomeric with some of the exceptions that are found to be dimeric like TyrRS, TrpRS and MetRS. They even considered charging the bulkier amino acids which are due to their open pocket binding site.

This class of synthetases is said to approach the tRNA acceptor end from the aminoacylates at the 2'-OH of an adenosine nucleotide and from the minor groove. Even though, many of the members of the class are seen to be capable of aminoacylation of both the 2' and 3' OH group.

It is believed that for many of the members of this class is releasing the charged tRNA in the form of aminoacyl-tRNA which further act as a rate limiting step used for the process of translation. On the basis of the similarities that are available in the sequence of class I synthetases, they are further being divided into five subclasses. These classes are designated as Ia, Ib, Ic, Id, and Ie respectively.

In the above sub-classes, the enzymes canrecognize the chemically related amino acids. The members of the subclass Ia includes LeuRS, IleRS, ValRS, and MetRS, that is being recognized as a hydrophobic amino acid having aliphatic groups

Subclass Ib synthetase includes GluRS, GlnRS and CysRS, these are considered to charge the amino acid as their substrate. They are considered to bind to their cognate tRNA before they experience the reaction of aminoacylation.

Subclass Ic synthetase includes TrpRS and TyrRS, in this, both of them are helping in recognizing the aromatic amino acid as their cognate substrate. The subclass Id is considered to include only ArgRS, because of its structural dissimilarity that happens with other sub-classes.

Subclass Ie synthetases are considered that although it has a structural similarity with the AARSs subclass Ib, still it has a unique -helix cage that makes it different from the other subclasses. These AARSs is considered to share several other common domains, these domains are like tRNA stem contact fold which is also known as SC fold, α helical like anticodon binding domain and connective to the peptide CP1 and CP2.

CP1 is found to be located at eh end of the first half of the strand D in the Rossmann fold; whereas, CP2 is found to be located in between the strand D and at the beginning of the second half of the Rossmann fold. CP1 is found to be slightly enlarged in Va1RS, LeuRS, and IleRS, then the MetRS and CysRS to favor the post- transfer editing activities of the mischarged tRNAs.

Va1RS, LeuRS, and IleRS, is considered that they are ranging from about 250 to 275 amino acid residues; whereas, MetRS and CysRS are considered to range from 100 to 50 amino acid, respectively.

Class II enzyme is sharing three degenerate motifs specifically in their active sites which is made up of anti-parallel β sheets. Motif 1 is considered to made up of an α helix that is further followed by a distorted β strands and they even containing a conserved pro residue that is playing an important role in the homo-dimerization.

Motif 2 and 3 are considered to contain a strictly conserved Arg residue that is found to be involved in the ATP binding process. There are approximately 250 amino acid residues that are present at the active site of class II enzyme. Class II AARSs are seen to approach the tRNA acceptor end from the major groove and they even aminoacylate the 3'-OH group of adenosine nucleotide. The sub-class II is considered to consist of ProRS, GlyRS, ThrRS, SerRS and HisRS. The aminoacylation domains that are present in this subclass is considered to be in the amino -terminal.

The subclass IIb is containing AsnRS, LysRS and AspRS, which are similar in structure to the C-terminal aminoacylation domain. The subclass IIc is consisting of PheRS, PylRS, GlyRS and AlaRS. Except for Py1RS, all the other members of this sub-class are existing in the tetrameric conformation. Whereas, AARSswith the subclasses IIa and IIb are considered to be dimeric.

3.5. AMINOACYL-TRNA SYNTHETASE REACTIONS

Modern protein aaRS is considered to achieve specific aa-tRNA synthesis in several steps. It is considered that their relative contribution towards specificity is said to vary from enzyme to enzyme.

3.5.1. Activation

In the first step, aaRS gets bind with their amino acid selectively and then they oppose its carboxyl for attacking at the α-phosphate of ATP.

aa + ATP + aaRS → (aa-AMP) aaRS + PPi

ATP gets bound for this particular amino acid activation reaction with the conformation that for the protein type, 5 oligo-phosphate often extends for the type I enzyme (Perona et al., 1993). They even bent back to the base for type II (Cusack et al., 1990).

Therefore, the chemistry is seemed to appear similar to the in-line displacement of the pyrophosphate from the ATP by the amino acid carboxyl oxygen. The catalytic role of type I and type II protein is seemed to appear similar.

Both these type acts simply: entropically for opposing tightly the ATP and amino acids by making many bonds to these substrates that are specific by using groups within the conserved type I and type II motifs.

Additionally, the conserved elements in both the type I and type II, aaRS seemed to offer the polar groups that help specifically for stabilizing the pyrophosphate leaving the group and the transition state that is used for the formation of the adenylate (aa-AMP).

Both the type of aaRS are considered to be prominently used by both the divalent metals and polar amino acids like lysine, histidine and arginine. Within each of these, the active sites superpose very well. Event the structural elements are considered to play the same role those are easily identified.

3.5.2. Pretransfer Editing

An amino acid is activated by the synthesis of it adenylate which is not committed necessarily to the aa-tRNA formation. There are some of the synthetases, primarily type I synthetases that are considered to have an editing function that helps in hydrolyzing the wrongly activated amino acid adenylate before it gets transferred to tRNA (Fersht, 1977).

$$\text{(aa-AMP) aaRS} \rightarrow \text{aa} + \text{AMP} + \text{aaRS}$$

Whereas, these specific hydrolytic discards of non-cognate adenylate is considered that it may require the presence of concurrent of the cognate tRNA, which is indicated by the tRNA in the scheme. It might be possible that tRNA is not reactant (Fersht 1977; Lin et al. 1984). It may act as a conformational effector.

The aminoacylation of the tRNA is not needed to be required step in the pre-transfer editing. The role of tRNA may be taken up by the DNA oligomers, those are selected for the affinity to synthetase (Hale and Schimmel 1996).

3.5.3. Aminoacyl Transfer

Aminoacyl-tRNAs are seen to arise from the attack of a terminal 2’ (3’) ribose hydroxyl on the carbonyl carbon of the aaRS-bound aa-AMP (Perona et al. 1993; Cavarelli et al., 1994).

$$\text{(aa-AMP) aaRS} + \text{tRNA} \rightarrow \text{aa-tRNA} + \text{aaRS} + \text{AMP}$$

3.5.4. Post-Transfer Editing

Even after the aminoacyl transfer process, aa-tRNA is not found to be necessarily committed with the ribosomal fate. The hydrolytic editing that formed aa-tRNA mistakenly (Eldred and Schimmel 1972; Yarus 1972) before it gets dissociated from the aaRS. aaRSis considered to clean up the errors that have not been found earlier.

The editing that is done at the last stage can either found to be large (Fersht and Kaethner, 1976) or can be small (Freist et al., 1996) factors in the overall accuracy. It depends on the aaRS, that the particular non-cognate amino acid or even from the conditions of the reaction (Freist, 1989).

(aa-tRNA) aaRS → aa + tRNA + aaRS

In order to carry out related activities, aaRS must be a complex catalyst. It is likely to have an interaction that occurs among the amino acid, RNA substrate and ATP of the aaRS. It is expected that from the first principles when there are three sites and these sites must be covering within the chemical bond length.

For example, the mobility of the ATP and Amino acid sites is considered to be implicated by the activation (Schmitt et al., 1994). The amino acid is being known for its capability of interacting with the tRNA site (Yarus and Berg, 1969). Further, these tRNA sites get interacted with the amino acid selection (Yarus and Berg 1969; Ibba et al. 1996).

Therefore, the interaction that takes place between the tRNA and amino acid sites are considered as sufficiently far- reaching, where the amino acids are not activated in a way that it forms adenylate in the absence of tRNA, as for Gln-, Glu-, and Arg-RS (Schimmel and Söll, 1979).

Therefore, the above succession of specific steps helps in giving the overall discrimination against the non-cognate amino acids. That is found to be in the range about 200-folds to 500,000-folds typically with the values of few thousand folds, as measured *in vitro* by the specific aa-tRNA synthesis (Freist et al., 1996).

Therefore, this spectrum of the specification is taken as the biochemical signature of a biologically fit aaRS activity, which is found to be capable of serving a functional translation system.

3.5.5. Self-Aminoacylating RNAs

Self - Aminoacylating RNAs are considered as a purified form from the pools of RNAs with the internal randomized sequence by simply selection and amplification process that take place *in vitro* and through cloning (Illangasekare et al., 1995).

After the process of incubation with the adenylate, aminoacylate RNAs are further made distinctly hydrophobic by the derivatization of their unique alpha-amino group. There is a majority of self-acylated aa-RNAs thoset are now hydrophobic and were resolved by using the reverse phase HPLC.

Phe-AMP + RNA → Phe-RNA + AMP

Phe-RNA + naphthoxyacetyl-NHS → naphthoxyacetyl-Phe-RNA + NHS

In the above equation, NHS is N-hydroxy succinimide.

Self-aminoacylation is considered as an interest that is outside the context of translation because it is among the first examples of the RNA catalysis at carbon. They even differ from the classic ribosome that is considered to perform phosphate chemistry.

The catalysis of self-aminoacylation shows that RNAs can help in stabilizing the tetrahedral, polar transition state which is considered as a characteristic of nucleophilic attack at the carbonyl carbon. Such type of transition state is being shared by an important class of reactions that is including the attack of amino groups at the acyl carbon.

For example, the peptide bond synthesis is being catalyzed by an rRNA rich sub-ribosomal particle (Noller et al., 1992). Even though the reaction is being found to be dependent on the protein (Noller, 1993). However, it is noticed that the argument specific for the ribozyme catalyzed peptide synthesis is being supported by the other evidence.

The establishment of the independent selections for the catalysis of pure RNA is considered to be similar to an aminoacyl transfer reaction (Lohse and Szostak, 1996; Zhang and Cech, 1997). Moreover, the large subunit rRNA is found to probably binding with both A and P site reactants within the few bond lengths for the formation of a peptide bond.

It is considered that a peptidyl transferase transition state analog is binding with the ribo-oligonucleotide which is containing the conserved sequence from the rRNA peptidyl transferase loop (Welch et al., 1997).

When there is hydrolysis, i.e., attack done by OH- or HOH at the carbonyl carbon, it leads towards similar transition state, which is also weakly catalyzed (with 10- to 100- folds stimulation that takes place over spontaneous), by the unselected RNAs which is acting on the esters (Piccirilli et al., 1992) and on the amide bonds (Dai et al. 1995, 1996).

Based on these examples where RNAs seem to host a similar tetrahedral transition intermediate, whereas other ribozymes are considered carrying out the related reaction at the carbonyl carbon which is likely to be found.

3.5.6. Six Metalloribozymes

From the above section, there have been six independently derived types of self-aminoacylating sequences that have been recovered (Illangasekare et al. 1997). All these sequences are found to be sensitive towards the periodate oxidation, which is a specific modification of the ribose hydroxyls.

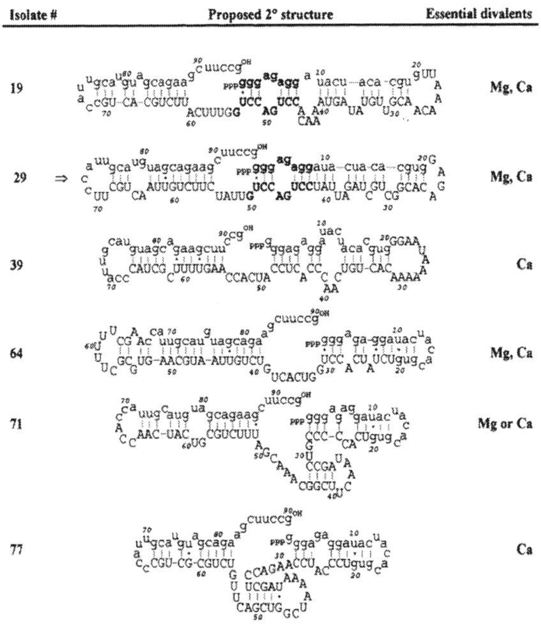

Figure 3.4. The six independently derived families of self-aminoacylating RNAs

Source: Image by http://rna.cshl.edu/

Therefore, all are considered to be analogous to the natural aa-tRNA that are forming terminal 2' (3') aminoacyl-ribose which is proved to be most prevalent family of sequences RNA 19, 29 (Illangasekare et al. 1995) and for the RNA 77 (M. Illangasekare and M. Yarus, unpubl).

This product is found to be chemically similar to the biological aa-tRNA products of the protein aaRS. Five of six types of RNA are considered to require calcium, and the other three of them also require magnesium with the calcium.Whereas, one RNA sequence is considered to be functional with either calcium or magnesium alone. Therefore, they are considered as metalloribozymes with calcium having a role that is uniquely prominent in the aminoacylation reaction (Illangasekare et al., 1997).

3.5.7. Reaction

The acylation velocity of RNA #29 is considered to respond linearly at the low concentration of Phe-AMP, but it can even observe to saturate at the high concentration of Phe-AMP. While implying a Michaelis constant KM = 9 and kcat = 1.2 min^{-1}.

The rate of aa-RNA formation is considered to be proportional to the hydroxyl ion concentration, which is found to be consistent with the hypothesis that, ionization of a group with the high pK value, a ribose hydroxyl which attacks the amino acid carbonyl is considered as an essential step in the reaction (Illangasekare and Yarus, 1997).

RNA-G-2' (3')OH → RNA-G-O^- + NH_3^+ ()CH-CO-AMP → RNA-G-O-Phe + AMP

This in turn implies that kcat represents the chemical step in the reaction. Considering that, if an amino acid is being removed then the RNA will repetitively catalyze its acylation with the rate which is undiminished, which indicates that it unaltered functionally by the acylation reaction.

3.5.8. Definition of the Active Site in the RNA 29 Family

Sequence comparisons can be used for identifying the required and dispensable features. For example, the linking sequence which is present between the two helices is considered to vary from RNA 19 and 29. Specific substitution experiments confirm that the linking sequence can also be varied within the limits.

According to a similar substitute, which suggests that the extension requires the two terminal nucleotides -CG -OH which must be of specific bases, but the substitution of the extension should be elsewhere. Comparison of RNAs 19 and 29 also suggesting that the proximal part of the 5' bulged hairpin domains should be conserved and therefore, it is required for the reaction.

This is being confirmed by the truncation of the hairpin that takes place beyond the conserved region and after the replacement of these 36 nucleotides by a stabilized tetraloop (29tr5) given in the figure below. Therefore, the resulting molecule is considered to be as active as the 95-mer parent.

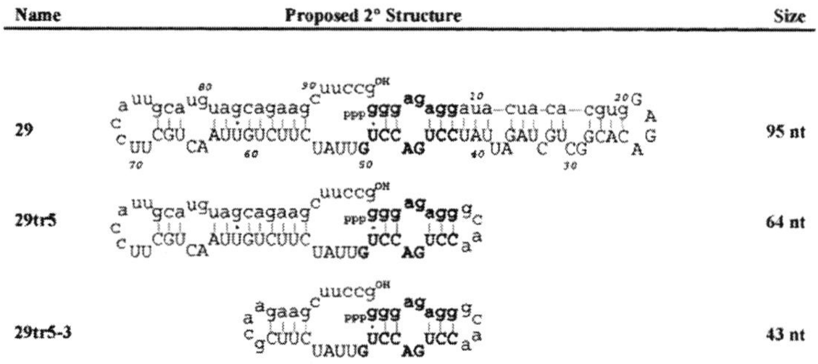

Figure 3.5. Truncation of RNA 29

Source: Image by http://rna.cshl.edu/

The 3' hairpin is found to be poorly conserved in the two independent molecules. It can be also abbreviated by the removal of 27 distal nucleotides which are being replaced by the tetraloop (RNA 29tr5-3). As a result, the doubly deleted 43-nucleotide transcript that will retain the full self-aminoacylation activity.

However, it is found that 3' domains can be radically truncated more evenly. It is not only conserved in the RNA 19 and 29, but there are modification interferences with the DEPC and DMS, that often suggests that even though the 5' domains are crucial, there is purine modification that is often innocuous in the 3' hairpin of 29tr5-3 (O. Kovalchuke and M. Yarus, unpubl).

Therefore, the 3' hairpin structure is entirely deleted. The presence of three nucleotides that are boldface in the RNA 29tr5-3-3 can be thought of as the bracketing nucleotides of the 3'hairpin. The central nucleotide replaces the 3' hairpin.

When these three positions are being randomized, then the molecules having various sequences, especially 3 of 4 nucleotides at each randomized position, were found to be active (M. Illangasekare and M. Yarus, unpubl).

The 3' hairpin is being replaced in these 32 nucleotide molecules in effect by the single nucleotide. The molecules which are shown are considered to be among the most active and are about 1/5th as active as the parental sequences.

This, as a result, reflects the primarily decrease in the kcat, which may be due to the slower chemical step and difficulty while fixing the 3' end. The resulting 32-mer is a self aminoacylating catalyst RNA 29tr5-3-3 which retains the essential magnesium and calcium sites. It is also due to the Michaelis complex with the Phe-AMP and the dependence on the value of pH that is considered as a characteristic of the selected 95-mer.

Despite deleting the 63 nucleotides which is approximately 2/3rds of the initial sequence, in this molecule the active site often appears to be present as virtually unaltered. The above Figure combines and superposes all the data which is published and unpublished, on the nucleotide requirements within the 32-mer catalyst.

By using a few of the nucleotides, it is possible to assemble a trans-aminoacylation active site. As it is important to bring the active site to bring the reactant together precisely, it is found to be most likely assembled within the conserved molecular environment in the RNA 29tr5-3-3, which is within the 5' bulge helix.

It is the place where the required 3' end can be packed into the bulge loop; this is done along with the molecule of the Phe-AMP and with the required mg++ and Ca++ (c.f. Illangasekare et al. 1997).

3.5.9. Amino Acid Specificity

The above- discussed RNAs of 29 families is being accelerated by their reaction more than 10^5-fold, but they are not amino acid specifically. The second- order rate is considered to be constant for the reaction of RNA and aa-AMP which is at the sub-saturating concentrations which are usually measured the characterize reaction velocity.

RNA 29 is a major substrate preferences being paradoxical, the reaction is considered to be few-fold faster with the Ser-AMP and Ala-AMP than with the Phe-AMP which is used to select them initially (Illangasekare et al. 1995).

It is noticed that they bind with the Phe-AMO, but they are being inhibited by the AMPwith the $K_I^{AMP} \approx K_m^{Phe\text{-}AMP}$, but it is not inhibited detectably by the free phenylalanine. The data that is taken together often suggests that the adenylate substrate can be treated as nonspecifically. This is so because it is found to be bound almost exclusively via AMP moiety.

Because of the amino acid specificity, it is considered as one of the hallmarks of the protein synthesis aaRS. It is even necessary for the accurate translation, amino acid specificity will be considered for the catalysis of aa-RNA synthesis that would be of interest.

It is considered that the RNA binding sites that are distinguishing the amino side chain are existing naturally (Yarus, 1988). These RNA binding can also be selected (Yarus, 1998), which is suggesting that self-acylating RNA can also exist via the possession of the specific amino acid site.

The amino acid side chain specificity is being observed for the RNA 77, for the purpose of selecting a Ca-specific ribozyme that prefers the Phe-AMP used for the process of selection. RNA 77 is considered to have the same 2' (3') terminal Phe-RNA product (M. Illangasekare and M. Yarus, unpubl). But it is considered to react about 900 -folds more quickly with the Phe-AMP than the derivatives of the RNA 29. Usage of an upper limit of 10^{-1} M^{-1} min^{-1} for the background acylation of randomized RNA by the Phe-AMP (Illangasekare et al. 1995).

RNA 77 is considered to accelerate its reaction for more than 7.3 X 10^7 folds which is basically under its optimal conditions. Despite an exceptional acceleration of the Phe-RNA production, RNA 77 is even believed to be less reactive with the Ala-AMP and Ser-AMP than is the RNA 29 family.

The second -order rate is constant $\approx k_{cat}/K_M$, and the rates of simultaneous reactions with the competing substrates are considered to be in the ration of their k_{cat}/K_M (Fersht, 1985). Therefore, the RNA 77 is being selected about 4000 -folds against the alanine which is basically based on rate and 20,000 -fold against the serine.

Also, there is another dimension towards RNA 77 specificity, therefore, only a minority of RNA 77 is said to appear to react with the Ala-AMP and Ser-AMP. Similarly, the same RNA 77 is seen in the presence of Phe-AMP which is fast reacting, kinetically homogenous and its reaction is also found to be almost complete is similar to that what is seen before the RNA 29 family (Illangasekare and Yarus, 1997).

In order to have a fast reaction of RNA 77 for yielding the Phe-RNA which is not a result of an exceptionally low KM, but it can be instead a fast k_{cat} which is estimated at hundred min^{-1}. Rate is considered to be proportionalto OH-, which means that the rate reflects the chemical step (M. Illangasekare and M. Yarus, unpubl).

Therefore, the superiority of the RNA 77 reaction depends on the fast and accurate transfer of the aminoacyl group, instead of being on the selective binding of the adenylates.

3.6. STRUCTURAL AND BIOCHEMICAL DISSECTION OF AMINO ACID ACTIVATION

The first structure of AARS is being solved at the high resolution; this was on the *Bacillus stearothermophilus* TyrRS. This structure was used as the basis for an exhaustive series of protein engineering studies that are being provided by the considerable insight into the mechanism that is being used for recognition of amino acid and activation of class I aminoacyl-tRNA synthetases.

Therefore, these experiments show that these reactions do not involve covalent or acid base catalysis. But instead of that, they do use the binding energy that helps in providing the enzyme reactant interactions in order to stabilize the process of translation state.

It has been noticed that the mutagenesis of the KMSKS motif in the TrpRS, MetRS and TyrRS, has shown that these residues are considered as a part of a mobile loop that helps in stabilizing the transition state of the reaction of the amino acid activation. This activation process uses ATP binding energy.

It is being shown that the HIGH – like motifs are playing a similar role in the TrpRS and MetRS. The detailed structure of the investigations of TrpRS has shown that the enzymes are assuming three distinct conformations during the tryptophan activation. These three conformations are: a closed pre- transition state ternary complex, an open ligand free form and a relaxed tryptophan adenylate product complex.

The structural changes are found to be associated with these various conformations that are being brought about by the movement of a small domain that includes the KMSKS and HIGH like motifs.

The role of KMSKS and the HIGH motif in the process of ATP binding and stabilization of the transition state by the GlnRS is found to be homologous from the structural studies, despite a notable difference which takes place in the reaction for the amino acid activation with this enzyme.

GlnRS together with the GluRS, LysRS I and ArgRS, is considered to require in the presence of the tRNA for the catalysis of the amino acid activation, which is in contrast with all the other synthetases and it is also considered that it does not require the tRNA.

One of the significant consequences of this requirement is that the sequence of tRNA is itself considered to strongly influence the affinity of the GlnRS for glutamine. Studies about the crystal structure of the GlnRS get complexed with the tRNA Gln and a glutaminyl-adenylate analog indicates that the results.

This is related to that, which is achieved from the terminal adenosine packing against, orienting the two aromatic residues which is in the glutamine binding site, a conclusion subsequently which is confirmed by the site- directed mutagenesis.

These various studies are providing a detailed understanding at the molecular level about the aminoacyl-adenylate synthesis by the class I AARSs. The depth about this knowledge is being illustrated by the success of recent attempts towards the engineer human GlnRS to accept the glutamic acid as a substrate.

The rapid advances that take place in the structural analysis of the class II AARSs have provided a detailed picture of the reaction that is important for the amino acid activation. The amino acid binding sites are considered to be rigid templates that are found to bind with the amino acid and ATP substrates.

This is in the optimal positions that take place for the transition of state formation by the in-line nucleophilic displacement. Within these binding sites, the specificity of the recognition is being done by the idiosyncratic contact to the side chain of the amino acid.

This process takes place in the context of an active site which shapes to accommodate the specific amino acids. The enhancement of the specificity of the substrate recognition is done by an induced fit mechanism. In this mechanism, the ATP binding leads towards the ordering of the motif 2 loops.

The importance of this substrate is to induce the ordering of the active site that varies for some of the Class IIAARSs. For example, SerRS, AspRS, in these AARSs the effect is mostly local, whereas in others like LysRS, HisRS, and ProRSare considered being a more global change that involves the relative position of the domain.

3.7. CONCLUSION

The process of aminoacyl-tRNA synthetase is considered as the heart of the modern translation which catalyzes the accurate biosynthesis of the aminoacyl-tRNAs. However, the first catalyst that is being made was aa-RNA for the coded protein synthesis which probably appears long before any other protein was introduced, that serves a pre-existing translation system.

Aminoacyl-tRNA is the substrates that are used for the translation and are pivotal in determining how the genetic code is being interpreted as an amino acid. The function of the aminoacyl-tRNA is to precisely match the amino acids with the tRNAs which are containing the corresponding tRNA by aminoacyl-tRNA synthetase.

Even though, the intrinsic proofreading and extrinsic editing are also considered as essential in several cases. According to the recent studies about the aminoacyl-tRNA synthesis is mainly prompted by the advent of whole genome sequencing.

Different classifications are also available for the aminoacyl-tRNA that plays an important role. The reaction of aminoacyl-tRNA takes place in several processes like activation, pre-transfer editing, post -transfer editing, self aminoacylating RNA and likewise many other processes are being included.

REFERENCE

1. Ibba, M. and Soll, D., 2000. *AMINOACYL-TRNA SYNTHESIS*. [ebook] Available at: <https://www.annualreviews.org/doi/pdf/10.1146/annurev.biochem.69.1.617> [Accessed 29 April 2020].
2. Marcey, D., 2020. *Trna Structure/Trna Synthetases*. [online] Earth.callutheran.edu. Available at: <https://earth.callutheran.edu/Academic_Programs/Departments/BioDev/omm/jsmolnew/trna/trna.html> [Accessed 29 April 2020].
3. n.d. *RNA Structure*. [ebook] Available at: <http://www.nsm.buffalo.edu/~koudelka/lecture_8_tRNA%20structure.pdf> [Accessed 29 April 2020].
4. Nguyen, H., Hoffer, E. and Dunham, C., 2019. Importance of a tRNA anticodon loop modification and a conserved, noncanonical anticodon stem pairing in tRNACGGPro for decoding. *Journal of Biological Chemistry*, [online] 294(14), pp.5281-5291. Available at: <https://www.jbc.org/content/early/2019/02/19/jbc.RA119.007410.full.pdf> [Accessed 29 April 2020].
5. Pak, D., Root-Bernstein, R. and Burton, Z., 2017. tRNA structure and evolution and standardization to the three nucleotide genetic code. *Transcription*, [online] 8(4), pp.205-219. Available at: <https://www.ncbi.nlm.nih.gov/pmc/articles/PMC5574529/> [Accessed 29 April 2020].
6. Rajendran, V., Kalita, P., Shukla, H., Kumar, A. and Tripathi, T., 2018. Aminoacyl-tRNA synthetases: Structure, function, and drug discovery. *International Journal of Biological Macromolecules*, [online] 111, pp.400-414. Available at: <https://www.researchgate.net/publication/322129042_Aminoacyl-tRNA_synthetases_Structure_function_and_drug_discovery> [Accessed 29 April 2020].
7. Rich, A. and RajBhandary, U., 1976. *TRANSFER RNA: MOLECULAR STRUCTURE, SEQUENCE, AND PROPERTIES*. [ebook] Cambridge. Available at: <https://pdfs.semanticscholar.org/ebef/71cf07eacd9beeb9ac693ab948f65397d023.pdf> [Accessed 29 April 2020].
8. Tharp, J., Ehnbom, A. and Liu, W., 2017. tRNAPyl: Structure, function, and applications. *RNA Biology*, [online] 15(4-5), pp.441-452. Available at: <https://www.researchgate.net/publication/319286753_tRNAPyl_Structure_function_and_applications> [Accessed 29 April 2020].
9. Yarus and, M. and Illangasekare, M., 1999. *Aminoacyl-Trna Synthetases*

And Self-Acylating Ribozymes. 2nd ed. [ebook] Colorado. Available at: <http://rna.cshl.edu/content/free/chapters/07_rna_world_2nd.pdf> [Accessed 29 April 2020].

10. Arnez and Moras, 1997
11. Cavarelli et al., 1994
12. Cusack et al., 1990
13. Dai et al. 1995, 1996
14. Eldred and Schimmel, 1972
15. Fersht and Kaethner, 1976
16. Fersht, 1985
17. Freist et al., 1996
18. Freist, 1989
19. Goebel and Helinski, 1970
20. Haenni and co-workers, 1975
21. Hale and Schimmel, 1996
22. Hoagland et al., 1957
23. Ibba et al., 1996
24. Illangasekare and Yarus, 1997
25. Illangasekare et al. 1997
26. Jacobson and K. B., 1971
27. Lin et al., 1984
28. Lohse and Szostak, 1996
29. Mischke, D., Kloetzel, P., Schwochan, M. 1975
30. Noller et al., 1992
31. Noller, 1993
32. Perona et al., 1993
33. Piccirilli et al., 1992
34. Schimmel and Söll, 1979
35. Schmitt et al., 1994
36. Welch et al., 1997
37. Yarus and Berg, 1969
38. Yarus, 1988
39. Zhang and Cech, 1997

4

Polypeptide Chain Initiation, Elongation and Termination

CONTENTS

The translation is considered as the second segment of the leading principle of molecular biology. It explains the way genetic codes would be used to form amino acid chains. The genetic codes constituted within an mRNA molecule are supposed to be decoded to form the particular sequence of amino acids in a polypeptide chain. Though the chemistry of forming the peptide bonds would be relatively intelligible, the process that would lead to the potentiality of forming a peptide bond would be extremely intricate.

The mRNA is considered as the template for accurate inclusion of individual amino acids, yet, the process would involve both tRNAs and rRNAs. Activated amino acids would be transferred into the ribosomes by the tRNAs, which would be made up of rRNA and ribosomal proteins. The ribosome would be linked with the mRNA and enable accurate access of activated tRNAs as well as comprising of the required enzymatic activities for catalyzing the peptide bond formation.

Transcription is followed by a translation which would occur in the cytoplasm. Similar to transcription, it would have three stages: initiation, elongation and termination. The chapter would be revolving around the three stages of translation and would provide an in -depth knowledge of the mechanism involved in the polypeptide synthesis.

Additionally, the chapter would provide a comparative analysis of the process among the prokaryotes and the eukaryotes as well as highlight their similarities and differences in the process of translation.

4.1. INTRODUCTION

No one would ever imagine how antibiotics kill bacteria, for example, when a person has a sinus infection. Though every antibiotic has its own way of functioning, yet, there are few which would have a very simple process of attacking the bacterial cells. They would eliminate the efficiency of making new proteins.

Such antibiotics restrict the translation process. The process by which piece of molecular information, known as messenger RNA (mRNA), is read by a cell and further used to build a protein, is called translation. In a normal bacterial cell, translation is a constant phenomenon. It would be a similar process

Once certain antibiotics are taken (example, erythromycin), the antibiotic molecule would fuse with the main translation molecules inside the bacterial cells and "stall" them naturally. With no possibility of making proteins, the functioning of the bacteria would finally get terminated and it would ultimately die. Hence, infections get cured when treated with antibiotics.

Translation helps cells to remain alive and comprehending the way it works (so that it can be cut off with antibiotics), can be a preventive measure against bacterial infection. To sum it up, the process by which information contained in a messenger RNA (mRNA) molecule is used for synthesizing a protein, is known as translation.By using genetic codes an mRNA sequence could be read during translation. The genetic codes are an algorithm that would interpret the technique of translating an mRNA sequence for the 20-letter code of amnio acids, considered as the basic components of proteins.

The genetic code is combined with a three-letter sequence of nucleotides called codons. Every codon is supposed to be corresponding with a particular amino acid or stop signal. The structure where translation takes place is called the ribosome. It is also regarded as the place for protein synthesis. The ribosome consists of a subunit, both small and large, and is believed to be a complex molecule made from numerous ribosomal RNA molecules and a few proteins.The mRNA molecular translation through the ribosome takes place in three stages: initiation, elongation and termination. Initiation is the process in which the small subunit of a ribosome would link with the start of the mRNA sequence. Following which a transfer RNA (tRNA) molecule consisting of amino acid methionine gets attached to something that's known as the start codon of mRNA sequence. In every mRNA molecule, the start codon would consist of AUG sequence and methionine codes.

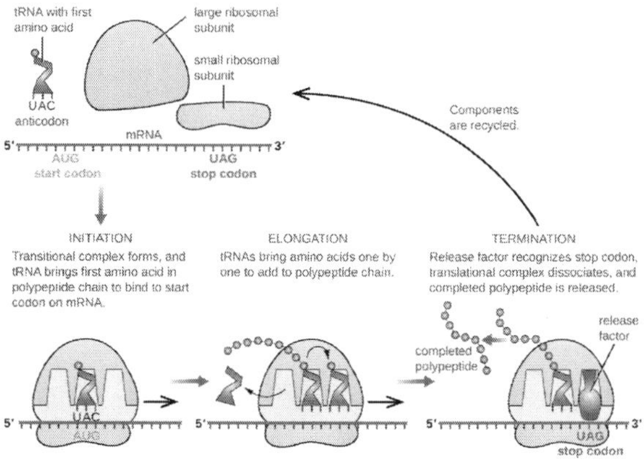

Figure 4.1. Stages of molecular translation

Source: Image by Wikimedia Commons

Further, the entire initiation complex is formed as a result of the binding of the large ribosomal subunit. The ribosome would keep translating every codon successively at the time of the elongation stage. Each equivalent amino acid is combined with the growing chain and connected through a bond called a peptide bond. Till the time each codon is read, the elongation process would continue.

The last stage would be for termination which would occur as the ribosome arrives at a stop codon (UAA, UAG and UGA). Without any tRNA molecules which could have identified such codons, the ribosome would consider the translation to be completed. It would be followed by releasing the new protein and thereby, bringing apart the translation complex.

4.2. POLYPEPTIDE CHAINS

Several kinds of molecules are used by the cells to carry out their daily activities. Proteins fall in the category of one of the most significant classes of molecules. They are the molecules that would have various functionalities within the cell. The functionalities would comprise of carrying things in and out of the cell, augmenting the chemical reactions and helping the cells to identify and bond to one another.

Hence, the versatile characteristics of the proteins make them useful for many purposes. For example, proteins could be associated with a swiss army knife that has multiple utilities. Similarly, the various structures present in proteins would define their exact functions.

There would be one or more polypeptide chains in proteins. Every polypeptide chain would have smaller sub-units or amino acids that would bind together. Amino acids are regarded as the building blocks of polypeptides whereas, polypeptides are considered to be the building blocks of proteins.

Considering a specific amino acid to be a paper clip, wherein a paper clip chain could be made by joining together many of the clips. Similarly, a polypeptide chain could be formed with the joining of several amino acids together. Moreover, a polypeptide chain itself can function as a protein. However, countless proteins are made up of numerous polypeptide chains.

As its already known that amino acids are considered to be the building blocks for proteins, and it would be comprised of an amino or NH_2 group and a carboxyl (acid) or COOH group. In proteins, there would be 20 distinct types of amino acids that are commonly found and for every protein molecule, there would be generally 300 or more amino acids.

Every amino acid would be different as per its "R" group. The molecular remainder is considered to be the "R" group of amino acid and the portion would not include the amino group, acid group and the central carbon. There would be a distinct "R" group for every particular amino acid. The specific chemical properties of every amino acid are dependent on the "R".

Amino acids, linked together by peptide bonds, where amino or NH_2 of particular amino acid would join with carboxyl (acid) or COOH group of any other amino acid, the formation of polypeptides and proteins would take place. In short, two or more amino acids bonded together by peptide bonds would be called a peptide.

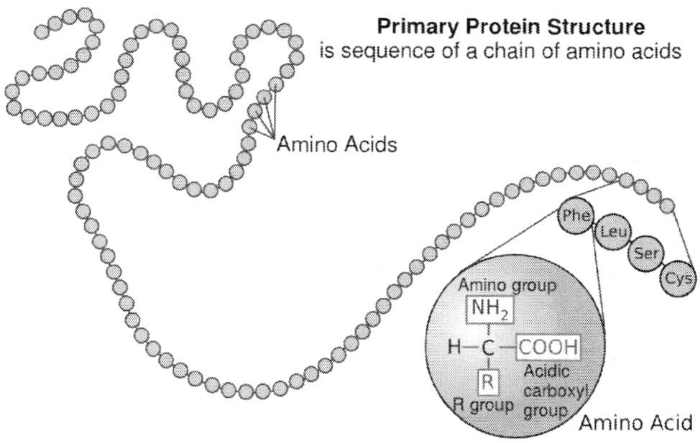

Figure 4.2. Primary Protein Structure

Source: Image by Wikimedia Commons

Whereas several amino acids bind together and form a chain, it is called a polypeptide. A protein would have one or even more polypeptides. Hence, amino acids joined together by peptide bonds and forming long chains are termed as proteins.

The real sequence of amino acids in the protein is termed as its primary structure and is well-defined by DNA. Further, the eukaryotes segregated into functional units known as genes. Deoxyribonucleotide bases forming a sequence adjoining a single strand of DNA that would be helpful in coding functional products are called genes. The functional products would include a particular messenger RNA molecule, transfer RNA molecule, or ribosomal RNA. Generally, the product would be messenger RNA (mRNA) and it would finally carry out the polypeptide or protein synthesis.

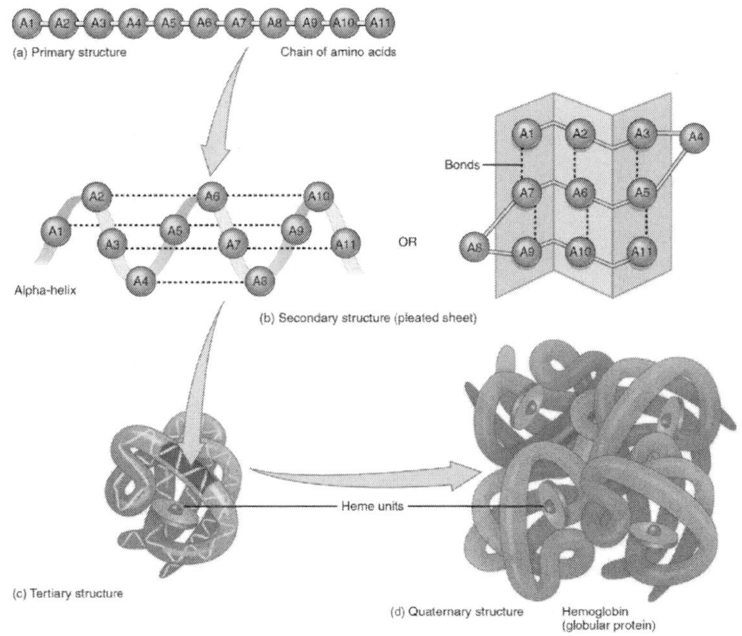

Figure 4.3. Peptide Bonds

Source: Image by Wikimedia Commons

Thus, it's often said that the sequence of deoxyribonucleotide bases within a gene would affirm the sequence of amino acid in a specific protein. Within the same protein few amino acids would be capable of interacting with other amino acids. It would ultimately help the primary structure to establish the final shape and consequently the protein's chemical and physical properties as well.

Hydrogen links that would form in between the oxygen atom of single amino acid along with the nitrogen atom of another result in the secondary structure of the protein. Thus, providing the two -dimensional pattern of an alpha-helix or a beta-pleated sheet to the protein or polypeptide.

The long chains formed in amino acids in globular proteins, like enzymes, would be pleated into a three-dimensional shape or tertiary structure. It is because few amino acids possessing sulfhydryl or SH groups would help in the formation of disulfide (S-S) bonds with different amino acids present on the same chain.

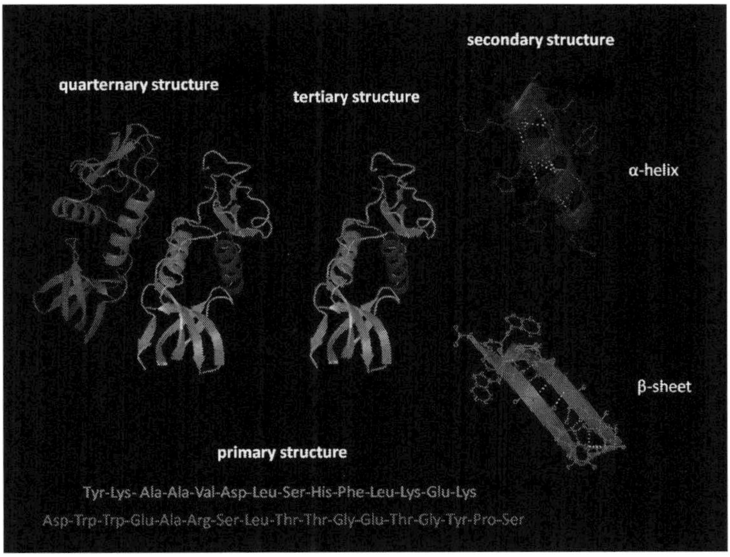

Figure 4.4. Protein Structure

Source: Image by Wikimedia Commons

Amino acids of the R group, like the hydrogen bonds, ionic bonds and hydrophobic interactions involving in other interactions, too, result in the tertiary structure. Few proteins, like the antibody molecules and hemoglobin, would create a quaternary structure by the joining of several polypeptides.

4.3. PREPARING POLYPEPTIDE CHAINS

A sequence of amino acids bonded together by peptide bonds would lead to the formation of a polypeptide chain. Every individual amino acid unit within a polypeptide is termed as a residue. The different ends of a polypeptide chain would maintain its polarity and that would include an α-amino group and an α-carboxyl group respectively at each end.

Traditionally, the amino end is considered to be the start of a polypeptide chain. Thus, in a polypeptide chain, the order of amino acids would be writing, beginning with amino- terminal residue. So, phenylalanine would be the amino-terminal (N-terminal) residue and leucine considered as the carboxyl-terminal (C-terminal in the pentapeptide Tyr-Gly-Gly-Phe-Leu (YGGFL). Possessing different chemical properties, Leu-Phe-Gly-Gly-Try (LFGGY) would be regarded as a separate pentapeptide.

The main chain or backbone would be the frequently repeated part and the individual side chains would form a variable part of a polypeptide chain. The hydrogen bonding capability in a polypeptide backbone would be quite strong. There would be a carbonyl group for every individual residue which is regarded as a decent hydrogen-bond acceptor in contrast to proline, which would be a part of NH_2 group, and an excellent hydrogen-bond donor. The interaction between such groups and other functional groups belonging to side chains would help in the stabilizing of specific structures.

The linear polypeptide chain in a few other proteins would be cross-linked. The oxidation of two cysteine residues leads to the formation of disulfide bonds, considered as the typical cross-links. The subsequent unit of linked cysteines would be known as cystine.

Often, many disulfide bonds would be present in extracellular proteins, whereas they might be usually absent in intracellular proteins. Seldom, would non-disulfide cross-links originated from alternate side chains, exist in certain proteins. For example, the collagen fibers present in connective tissues would be strengthened according to the specified way and so would the fibrin blood clots.

4.4. THE MECHANISM OF PROTEIN SYNTHESIS OR TRANSLATION

The genetic information existing in mRNA would manage the sequence of each amino acids and help in the formation of a polypeptide chain, at the time of translation. The translation would take place for both prokaryotes and eukaryotes. The basic process of translation would be same in case of both and members belonging to both the groups would be using information available in mRNA.

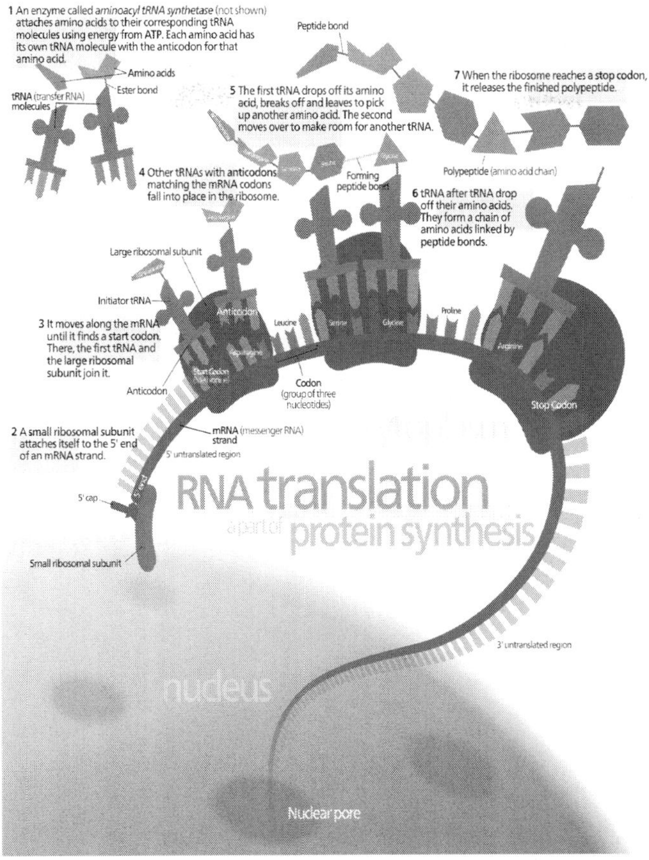

Figure 4.5. RNA translation

Source: Image by Wikipedia

The information of mRNA is supposed to have been derived from the transcription of DNA for synthesizing proteins while considering the ribosome to be the tool. The core steps leading to protein synthesis or translation could be categorized under three phases; initiation, elongation and termination.

4.4.1. Polypeptide Chain Initiation

The start of protein synthesis would occur by forming an initiation complex. The complex, in case of an *E.coli*, or prokaryote, would comprise of the small 30S ribosome, the mRNA template, three initiation factors (Ifs; IF-1, IF-2 and IF-3) and distinct tRNA initiator known as tRNA*fMet*. The tRNA initiator would connect with the start codon AUG (or sometimes, GUG), join with a formylated methionine termed fMet and could even get the IF-2 bonded.

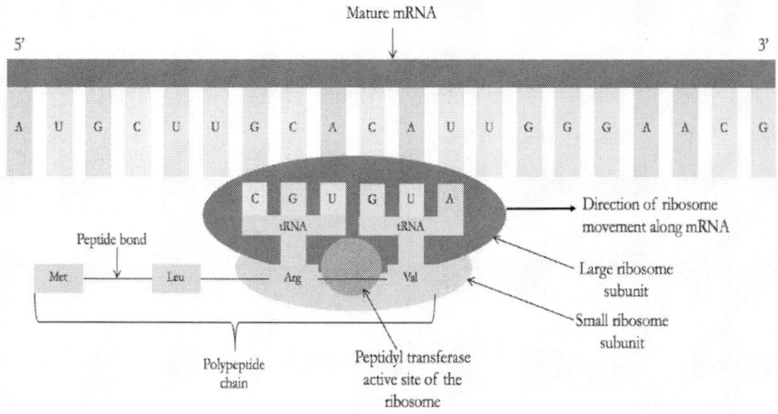

Figure 4.6. Translation of mRNA for Producing a Protein

Source: Image by Wikimedia Commons

While *E.coli* synthesizing each polypeptide chain in the initial stage, formylated methionine would be inserted through fMet−tRNAfMet, however, it would usually get detached after the completion of the translation process. A non-formylated methionine would be inserted through a standard Met-tRNA^{Met} at the time of translation elongation while encountering an in-frame AUG.

An upstream order of the first AUG codon in *E. coli*, known as the Shine-Dalgarno sequence (AGGAGG), would interact with the rRNA molecules comprising the ribosome. 30S ribosomal subunits would be bonded through such interactions at the exact location on the mRNA template. At the time of translation, occurring at the beginning of elongation as well as during translocation of the ribosome, Guanosine triphosphate (GTP), a type of purine nucleotide triphosphate, would act as a medium of energy.

An equivalent initiation complex would be formed in eukaryotes, consisting of mRNA, the 40S small ribosomal subunit, IFs and nucleoside triphosphates (GTP and ATP). In eukaryotes, the charged tRNA initiator known as the Met-tRNAi would not bond with fMet. However, it would be different from other Met-tRNAs where it could join IFs.

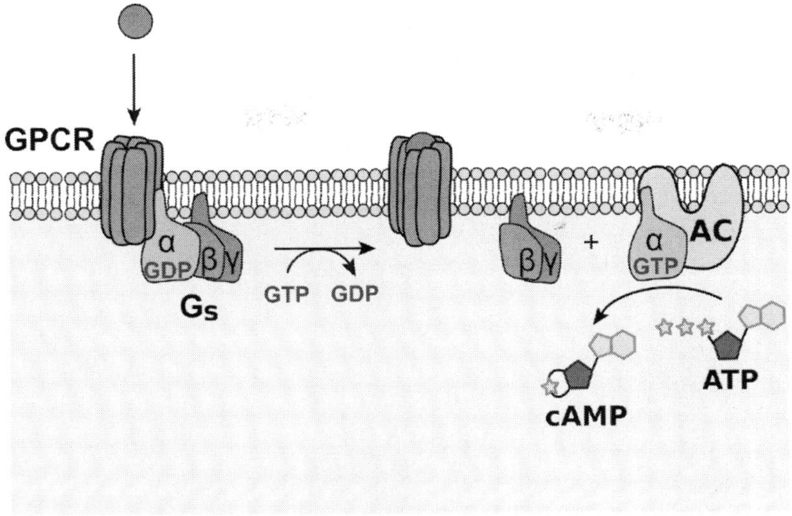

Figure 4.7. Nucleoside Triphosphate

Source: Image by Wikipedia

The eukaryotic initiation complex would identify the 7-methylguanosine cap at the 5′ end of the mRNA, rather than leaving it at the Shine-Dalgarno order. The ribosome 5′ cap movement would be supported by cap-binding protein (CBP) and various other IFs. On reaching the cap, the initiation complex would record collectively the mRNA in 5′ to 3′ direction looking out for the AUG start codon.

Several eukaryotic mRNAs would be translated from the first AUG, yet it would not be applicable in all situations. As per the rules of Kozak, the nucleotides encompassing the AUG would show if it's the exact start codon. The rules of Kozak reveal that the resulting consensus order needs to be present around the AUG of vertebrate genes: 5′-GCCRCCAUGG-3′.

The R (in case of purine) points towards a location that could be either A or G, but shall not be C or U. most importantly, the nearer the sequence would be to the consensus, the greater the translation proficiency. On recognizing the suitable AUG, the other proteins and CBP would get detached and the 60S subunit would join with the Met-tRNAi complex, mRNA and the subunit 40S. It is considered as the final step that would complete the initiation translation in eukaryotes.

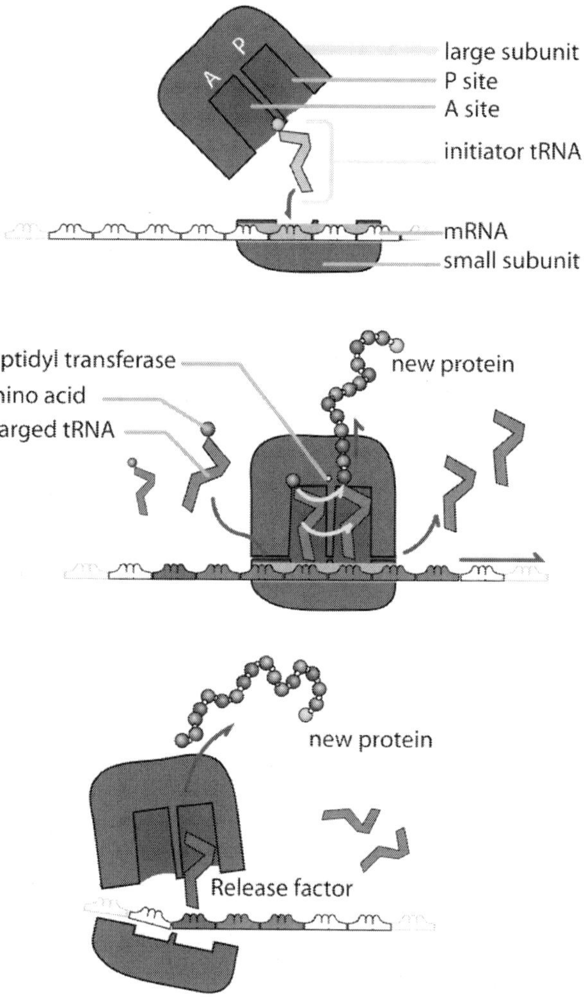

Figure 4.8. tRNA Ribosome

Source: Image by Wikimedia Commons

4.4.2. Characterization of Translational Initiation Between Eukaryotes and Prokaryotes

To start protein synthesis, a ribosome accompanied by its designated initiator methionyl-tRNA needs to be prepared at the start codon of an mRNA. The process would involve coordinated actions of three translation initiation factors (IF) in prokaryotes and a minimum of 12 translation initiation factors in eukaryotes (eIF). The eIF1A and eIF5B factors in eukaryotes would exhibit a vast resemblance of amino acid sequence with the IF1 and IF2 factors in prokaryotes.

Through the combination of two-hybrid, coimmuno-precipitation, and *in vitro* binding assays a direct interaction would be observed in eIF1A and eIF5B. Moreover, the binding location of sIF1A is considered to be linked to the C-terminal region from eIF5B. The particular region concerning eIF5B would be regarded as crucial for the process of growth *in vivo* and translation process *in vitro*.

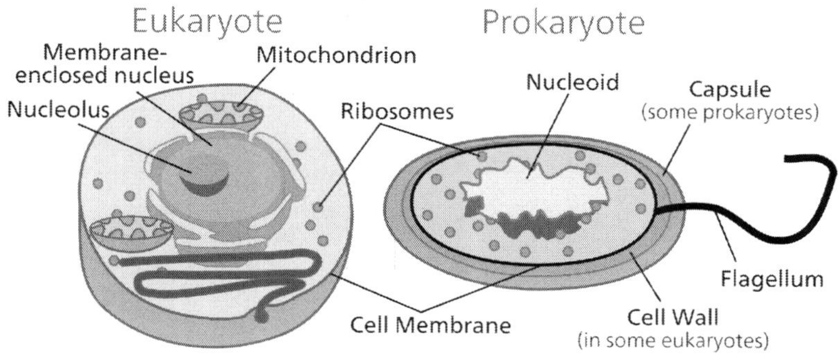

Figure 4.9. Eukaryote and Prokaryote Cell

Source: Image by Wikipedia

The slow-growth phenotype of yeast strains would be intensified by the overexpression of eIF1A, intimidating C-terminally to concise eIF5B. Such results signify the physical interaction between the progressively maintained eIF1A and eIF5B factors that would perform an important part in the translation initiation. It would possibly control or stabilize the binding of methionyl-tRNA with the ribosomal P site.

The basic process of translation initiation seemed to be maintained between prokaryotes and eukaryotes. The Met-tRNA initiator would be attached to the small ribosomal subunit and the complex would be confined to the AUG start codon of an mRNA. In prokaryotes, three translation initiation factors (IFs) are supposed to have been recognized. In the case of binding fMet-$tRNA_i^{Met}$ with the 30S ribosomal, the liability would be on subunit factor IF2.

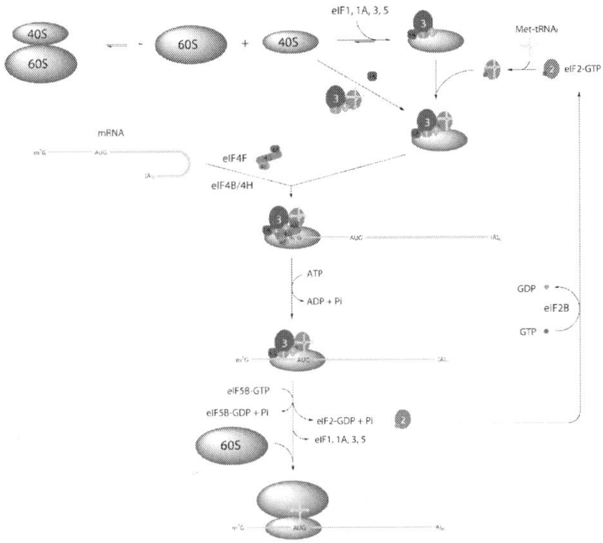

Figure 4.10. Eukaryotic initiation

Source: Image by Wikimedia Commons

The subunit 30S would be bonded by the factor IF1 which would also be responsible for the protection of the ribosome site (A site). The site would be the same as the elongation factor EF-Tu-GTP-aminoacyl-tRNA complex. Though a distinct function might not have been accredited to IF1, yet, it would be liable for promoting IF2 activities.

Ribosomal complexes would be separated by factor IF3, probably it would help to produce a team of small ribosomal subunits for the translation initiation process. Additionally, the process of ribosomal recycling is supposed to be newly implicated with IF3 and it would be followed by the translation termination process. The process of translation would be more intricate eukaryotes as compared to the prokaryotes. Hence, it might be expected that the number of initiation factors in eukaryotes would be greater.

The factors that facilitate the binding of Met-tRNA$_i^{Met}$ small 40S ribosomal subunit in eukaryotes would be factor eIF2. Factor eIF1A that was earlier known as eIF4C, is supposed to have been testified to operate in subunit dissociation along with the promotion and stabilization of Met-tRNA binding with the 40S subunit. Further, the toe-printing assays in the ribosome would determine the requirement of eIF1A for forming a 48S preinitiation complex. It would help to a 40S subunit accompanied by associated initiation factors and Met-tRNA$_i^{Met}$ would be bound on the

AUG start codon belonging to mRNA resulting in ribosomal scanning.The sequence resemblance in prokaryotic IF2 is considered to be quite prominent in translation initiation factor eIF5B. Earlier it was supposed to have been referred to as yIF2 or hIF2 in case of yeast (*Saccharomyces cerevisiae*) or human IF2, respectively. As per a recent report, eIF5B would play a crucial part in joining the subunit during protein synthesis.

Though eIF5B would be regarded unimportant in yeast, yet, if the FUN12 gene encoding eIF5B is removed, it would damage the translation initiation *in vivo* and *in vitro*. As a result, it would lead to extreme stagnant growth phenotype. However, overexpression of $tRNA_i^{Met}$ would help to moderately suppress the stagnant growth phenotype, recommending the importance of eIF5B for Met-$tRNA_i^{Met}$ binding to ribosomes.

In the case of yeast factor for promoting growth *in vivo* and translation initiation restoration in extracts made from fun12Δ strains, human eIF5B could act as a functional substitute. Similar to prokaryotic IF2, eIF5B would also include a consensus GTP-binding domain. The factor contained ribosome-reliant GTPase activity that would seem to be necessary for dissociation of the factor from 80S ribosomes resulting in joining of the subunit.

GTP binding by eIF5B is considered to be a regular significant role by which mutations in the GTP-binding domain of human eIF5B impedes translation in yeast and human cells. In a recent report, an interaction between *Drosophila melanogaster* eIF5B and the DEAD box RNA helicase VAS was observed, which was encoded by genevasa.

The functional interaction of proteins had been suggested by the genetic interactions between VAS and eIF5B mutations. Though, in translation the function of the said interaction is unknown. In Archaea, an eIF5B/IF2 homolog is supposed to have been identified, and it was observed that the protein could be capable of partially substituting the yeast protein in the case of both *in vivo* and *in vitro*.

Interestingly, an effective sequence similarity could be observed between archaeal and eukaryotic eIF1As and prokaryotic IF1. Therefore, it could be concluded that a pair of universally maintained translation initiation factors could be formed from eIF5B/IF2 and eIF1A/IF1.

Imitation of tRNA would be a general concept derived from the structural studies of translation factors. A comparative study was conducted between the crystal structures of the prokaryotic elongation factor EF-Tu–GTP–Phe-$tRNA^{Phe}$ ternary complex and elongation factor EF-G-GDP binary complex.

It had been observed that the III and IV domains of EF-G had a strong similarity between the anticodon stem and loop of the tRNA in the EF-G ternary complex.

Additionally, the GTP-binding domains of EF-Tu and EF-G could probably be applied in two structures. Provided that EF-Tu would bind the aminoacyl-tRNA to the ribosomal A region, the EF-G structure would maintain the idea of EF-G would get bonded to the ribosomal A region and help to promote translocation of the peptidyl-tRNA from the A site to the P site at the time of elongation.

Certain similarities would be in the structure of the translation termination factors from humans and that of a tRNA molecule, even though the tertiary structure of the supposed tRNA-imitating domains in eRF1 would not be similar to the corresponding domains in EF-G. Equally, the structural elements, including tRNA replicating domain of the bacterial ribosome recycling factor RRF would look neither like those of EF-G nor those of eRF1. Hence, protein would have established numerous ways to imitate the structure of tRNA.

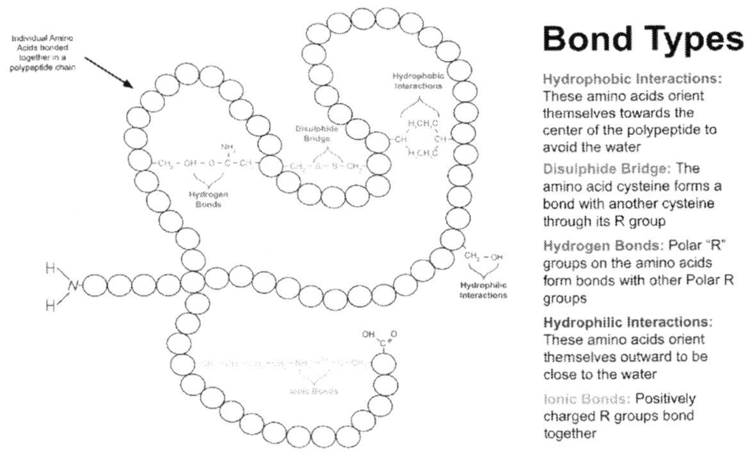

Figure 4.11. Tertiary Structure of Protein

Source: Image by Wikimedia Commons

In recent times, a detailed molecular imitation hypothesis to translation initiation was done by Brock et al. 1998. Based on amino acid sequence resemblance between various segments of the C-terminal regions of EF-G and prokaryotic IF1 or IF2, it was supposed to have been recommended that IF1 and IF2 would have a mutual interaction and together structurally imitate

EF-G. Since Met-tRNA$_i^{Met}$ would be considered to be directly delivering to the ribosomal P region, the idea of an IF1-IF2 complex structurally imitating EF-G and binding to the A site on the ribosome would be most likely.

As per the said paradigm, the binding of A site by the IF1-IF2 complex would sterically block the site and direct the Met-tRNA$_i^{Met}$ binding to the P site. The paradigm would be supported by the fact that IF1-IF2 binding is considered to have been localized to the ribosomal A site. The cross-linking studies of protein would suggest that IF1 and IF2 could interact if bonded with the 30S ribosomal subunit.

However, no direct interaction between the isolated proteins from the ribosome has yet been reported. The molecular imitation hypothesis for IF1 and IF2 fascinated researchers a lot. Hence, based on the extreme sequence conservation of eIF1A and eIF5B with IF1 and IF2, they tried to test the hypothesis related to the interaction factors of eukaryotes.

4.4.3. Polypeptide Chain Elongation

There are three compartments in the 50S ribosomal subunit of *E. coli*; the A (aminoacyl) site would bind the incoming charged aminoacyl tRNAs. The P(peptidyl) site would bind the charged tRNAs that carry amino acids forming peptide bonds with the expanding polypeptide chain, but not yet separated from their equivalent tRNA.

The E (exit) site would discharge the separated tRNAs to recharge them with free amino acids. It would lead to the formation of an initiation complex with a free A site prepared to receive tRNA in correspondence to the first codon after the AUG.

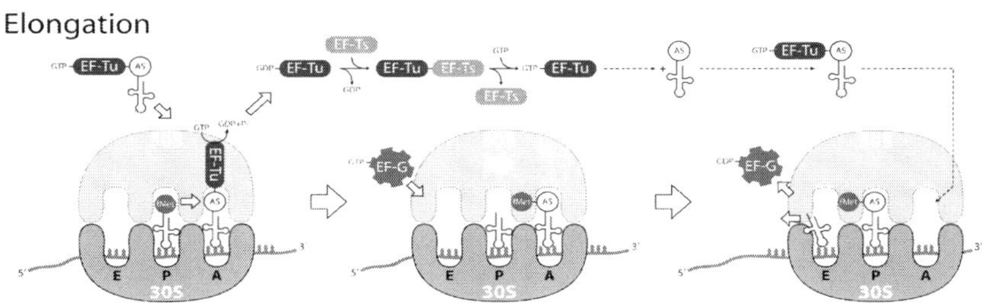

Figure 4.12. Polypeptide Chain Elongation

Source: Image by Wikimedia Commons

During protein synthesis, the universally well-maintained A (aminoacyl), P (peptidyl) and E (exit) sites of the mRNA programmed ribosome would be successively occupied by the tRNA.The elongation cycle of translation is considered to be a continuous three-step process that would be catalyzed by the ribosome and two GTPases, EF-Tu and EF-G (eEF1A and eEF2 in eukaryotes).

First, while decoding, EF-Tu would deliver an aminoacylated-tRNA (aa-tRNA) molecule as a segment of the ternary complex of aa-tRNA·EF-Tu·GTP to the A/T region. The orientation of the tRNA entering the ribosome would not be appropriate for codon-anticodon interaction. Moreover, the stem-loop of anti-codon in the tRNA would require to be strongly bent at the A/T region for allowing the anticodon to interact with mRNA.

While the aa-tRNA would be considered to be cognate – i.e., once its three-base anticodon would create complementary base pairs with messenger RNA codon – GTP would be hydrolyzed and the aa-tRNA,onEF-Tu·GDPgetting dissociated, would be accommodated in the ribosomal A site.

In the second step, the rRNA of the large subunit would catalyze the peptidyl-transferase reaction and it would occur instantly following the aa-tRNA accommodation. The nascent peptide chain would be transferred from the peptidyl-tRNA in the ribosomal P region to the contained aa-tRNA in the A site. It would result in the "pretranslocational" (PRE) state the ribosomal translation, along with the deacylated tRNA in the P region and a peptidyl-tRNA in the A region.

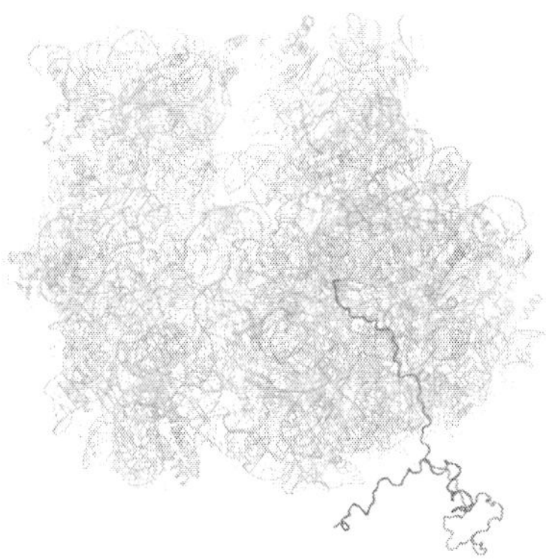

Figure 4.13. Ribosome – Nascent Chain Complex

Source: Image by Wikipedia

Translocation would be the final step of the elongation cycle. It would get the ribosome into the "posttranslational" (POST) state with an unoccupied A site, peptidyl-tRNA in the P site and deacylated tRNA in the E site. The binding of EF-G and following GTP hydrolysis would help to catalyze translocation. The tRNA molecules movement would be associated with that of the bound mRNA chain over three bases, letting the subsequent codon of mRNA in ribosomal A site as being presented for decoding.

Every step in the process of elongation cycle of translation, like decoding, peptidyl-transfer and translocation would put forward an enigma in itself. However, the technique of peptidyl-transfer would have received specific attention (Barta A, Dorner S, Polacek N, 2001) and in any case, a segment of the tRNA decoding selection step would have been explained in molecular detail (Ogle JM , Brodersen DE , Clemons WM, Jr , Tarry MJ , Carter AP, Ramakrishnan V, 2001). Further, the molecular methods of EF-G-mediated translocation would not have been comprehended better until recently. The major findings would comprise of the discovery of the tRNA hybrid states (Moazed D, Noller HF, 1989) and the outcome of kinetic experiments that GTP hydrolysis would lead to translocation (Rodnina MV, Savelsbergh A, Katunin VI, Wintermeyer W, 1997).

Vital cryo-EM findings pertinent to translocation would be the visualization of EF-G binding to the ribosome (Agrawal et al., 1998;Agrawal et al., 1999; Stark Het al., 2000); the t-RNA visualization in the P/E hybrid state (Valle et al., 2003; Gao et al., 2007; Taylor et al., 2007; Agrawal et al., 1999); finding the ratchet motion, the small vs. large subunit rotation upon binding of EF-G or eEF2 (Spahn et al., 2004) to the ribosome; and the latest one would be elucidating the GTP hydrolysis role in eEF2-induced eukaryotic ribosome translocation.

4.4.4. Translation Elongation Factors

Proteins are the translation elongation factors that would play two major roles at the time of elongation cycle of protein biosynthesis on ribosome.

First of all, elongation factors would be responsible for bringing aminoacyl-transfer RNA to the ribosome at the time of protein synthesis.

Secondly, in translocation an elongation factor would be involved, the elongation step during which the peptidyl-tRNA would move from one ribosome region to another as the messenger RNA is considered to be moving through the ribosome.

Both steps would be resulting in the hydrolysis of guanosine triphosphate (GTP). Moreover, the arrangement of the elongation factors would change based on whether they are bonded to the GTP or guanosine diphosphate. The structure and function would be similar in the elongation factors of archaea and bacteria (both being a type prokaryotes) and eukaryotes, like the protein biosynthesis steps wherein they would participate.

Currently, hundreds of protein modifications have been known and that would make proteomes much more complex than the encoding genomes could have predicted. The biological functions would be modulated or the subcellular location of the proteins would be altered by the covalent modifications. It would as well impact the protein interactions with a variety of molecules, like nucleic acids, lipids or other proteins (Walsh et al., 2005; Ahrne et al., 2010).

Specific modifications would be usually present on several proteins and frequently the proteins would carry various modifications at numerous amino acid residues (Yang, 2005). In protein modifications, the synthesis and attachment would usually involve various gene products and set of metabolites. It would make the events costly for a cell with respect to substrate and energy requirements.

However, additional functions for the proteins might be generated through the modifications, or novel pathways of regulation would be allowed provisioning a cell with additional means of diversification and development. Certain modifications would be brief and hence, would be dependent on quick attachment and removal of molecules from target proteins. There would be others, stable and attached to proteins soon after their synthesis or before degradation (Yang XJ, 2005).

Between several protein modifications, three would be regarded as exceptional wherein they would occur on single proteins: ethanolamine phosphoglycerol (EPG), diphthamide and hypusine. Especially, the corresponding proteins that would carry the modifications, factor 1A (eEF1A) and factor 2 (eEF2) in eukaryotic elongation as well as factor 5A (eEIF5A) in eukaryotic initiation, respectively, would be involved in the elongation steps of translation.

During translation, the elongation of polypeptide chains would be considered as a conserved process between prokaryotes and eukaryotes. Elongation with single steps would consist of (a) aminoacyl-tRNAs binding to the A(aminoacyl)-site of the ribosome, (b) forming peptide bond with the adjacent peptide-tRNA at the P(peptidyl)-site and (c) the extended peptide-tRNA translocating from the A-site to the P-site and the prior loaded tRNA from the P-site to the E(exit)-site.

Such steps would be properly conserved between organisms and enzymatic connection with ribosomal RNA at the transpeptidation center and are mostly accepted nowadays. Therefore, homologs of most factors that would have any involvement in elongation could be located across bacterial, archaeal and eukaryotic genomes.

eEF1A, eEF2 and eIF5A would be phylogenetically considered as one of the most extremely conserved proteins. During the elongation of translation, their biological roles would include the following: eEF1A (known as EF-Tu in bacteria and aEF1A in archaea), among the most substantial cytosolic proteins, would catalyze binding of aminoacyl-tRNAs to the A-site of the ribosome. Additionally, as per reports, it is supposed to have been involved in several other functions (like the moonlight functions).

On the contrary, eEF2 (known as EF-G in bacteria and aEF2 in archaea) would be involved in peptide-tRNA complex translocation from the A- to the P-site. Meanwhile, eIF5A (known as EF-P in bacteria and aIF5A in archaea) would direct stimulate protein elongation, though its exact mode of action on the ribosome would be vague (Saini et al., 2009). The proper situating of the initiator-tRNA-methionine complex would be facilitated by bacterial EF-P, at the P-site (Blaha et al., 2009).

eEF1A and eEF2 both would be regarded as GTP-binding proteins. It implies that their enzymatic activities would need the GTP to GTP hydrolysis. Interestingly, translation elongation, the GTPases involved would show a notable structural resemblance that would be indicating at a common ancestral GTPase (Hartman H & Smith TF, 2010).

Its apparent function was supposed to be the transportation of aminoacyl-tRNAs to an ancestral membrane-bound self-folding RNA which would get the peptide bond formation catalyzed and would constitute the original peptidyltransferase center, evolving later into the equivalent domain of the ribosomal large subunit.

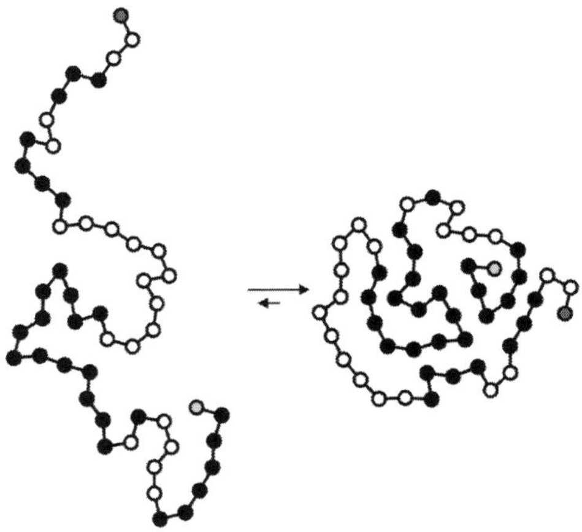

Figure 4.14. Protein Folding

Source: Image by Wikimedia Commons

The co-evolution of translational GTPases with ribosomal structures might have occurred for allowing interaction of GTPases with ribosomal structures by inserting new structural elements (Hartman H & Smith TF, 2010). According to the co-evolution concept between proteins and RNA structures, a notable molecular imitation between proteins and tRNAs would be shown by the elongation (and termination) factors of translation. For example, in EF-G from *Thermus thermophilus*, its crystal structure would accurately fit the structure of the ternary prokaryotic EF-Tu-GDPNP-Phe-$tRNA^{Phe}$ complex (Nakamura Y & Ito Km, 2003).

Further, the crystal structure of EF-P from *Escherichia coli* accompanied by its post-translational lysine modification similar to the covalently bound amino acid lysine charged to the 3′ end of a tRNA would imitate the structure of a charged tRNA (Yanagisawa et al., 2010).

The distinct modifications associated with eEF1A, eEF2 and eIF5A are considered to be known for decades. Besides, for production and attachment to proteins through their biosynthetic precursors and pathways are supposed to be partly established. Astonishingly, their biological functions are considered to be incomprehensible regardless of the fact that EPG, diphthamide and hypusine and vital proteins involved in an extremely conserved process. It implies elongation of proteins during translation and thoseexact species variants of the three proteins are supposed to be crystallized as well as their 3D structures solved.

Remarkably, despite the fact that apart from the function, even 3D structures of e(a)EF1A/EF-Tu, e(a)EF2/EF-G and e(a)IF5A/EF-P proteins are supposed to be conserved during evolution, there would be striking differences exhibited due to the presence of EPG, diphthamide and hypusine. However, attachment of hypusine to e(a)IF5A/EF-P proteins are considered to be established in all three domains of life, the modification in diphthamide seemed to be found in e(a)EF2 of eukaryotes and archaea, however, not in EF-G of bacteria, while information of EPG is supposed to have been found in eEF1A of eukaryotes.

As per researches, rare post-translational modifications of three varied translation elongation factors would be existing in all cells and involve in a conserved mechanistic route among eukaryotes and prokaryotes. To maintain the activity (and perhaps even the proper structure) of acceptor proteins, EPG, diphthamide and hypusine would be essential. The biological implications of such modifications might simply become obvious *in vivo* or due to certain stress or competitive circumstances, which until now, wouldn't have been imitated in the laboratory.

In the case of all eukaryotes, that have been studied so far, the eIF1A, eEF2 and eIF5A function would be essential for the survival of a cell. According to researches, cross-complementation experiments accompanied by paralog prokaryotic and eukaryotic factors are considered not being reported as of now. One of the probable reasons for such experiments not being successful would be as a consequence of co-evolution of such proteins with their interacting partners that might have induced subtle variations that would not support cross-complementation of single paralogs in various organisms.

It is indefinite whether EPG, diphthamide and hypusine would be responsible for protein-protein interactions. The existence of potential knockout/knockin and knockdown methods by the usage of mono- and multi-cellular organisms might permit us to know the facts about the importance of such modifications to be extended in the near future.

Many would anticipate the reason behind the three modifications EPG, diphthamide and hypusine being restricted to single proteins as well as the reason for the three modified proteins to be all involved in translation. Hence, it has been proposed to consider the modifications as the remnants of an evolutionary process which might have had been more common in an ancient world, i.e., that numerous proteins would have been modified by EPG, diphthamide and hypusine.

During the stages of evolution, nevertheless, such modifications might have mainly disappeared, leaving behind the translation elongation proteins e(a)EF1A/EF-Tu, e(a)EF2/EF-G and e(a)IF5A/EF-P. They would be extremely conserved between organisms and hence, EPG, diphthamide and hypusine might accomplish important functions for improving accuracy or catalytic activity of enzymes trying to interact with translating ribosomes. In the case of diphthamide and a part of hypusine, the functional essence is supposed to have been established. However, no such functional part has been reported for EPG.

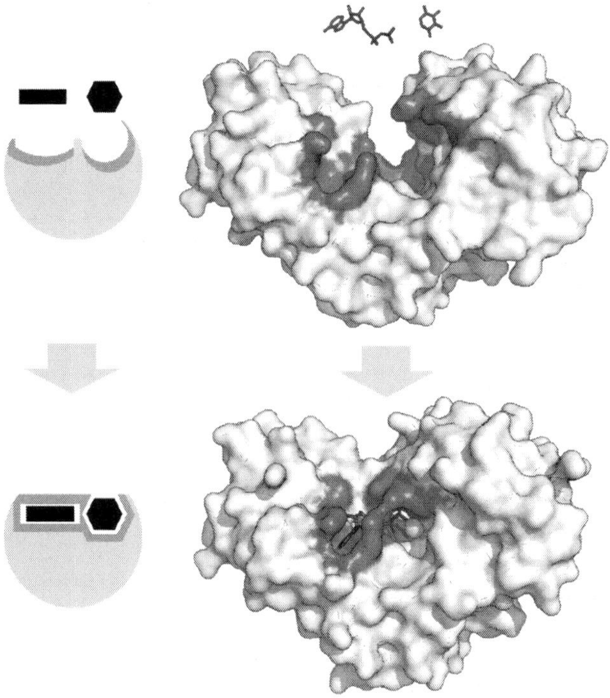

Figure 4.15. Enzymes

Source: Image by Wikipedia

4.4.5. Polypeptide Chain Termination

The protein synthesis termination would include the decoding of a stop signal by means of an interaction between RNA (rRNA and mRNA) and proteins (release factors RFs). It would enable the hydrolytic release of the nascent polypeptide chain from the peptidyl-transferase center of the ribosome. Though they would possess few common traits in their translation termination system, yet, prokaryotes and eukaryotes would exhibit major differences.

Bacteria would have two Class I decoding RFs (RF1 and RF2) accompanied by overlapping codon particularity, whereas, eukaryotes would have a single decoding factor, eRF1. Every individual prokaryotic factor would respond to UAA, while UAG would be decoded only by RF1 and UGA would be decoded solely by RF2. On the contrary, eRF1 would possess an absolute decoding capacity and can promote completed polypeptide released in response to either of the three stop codons.

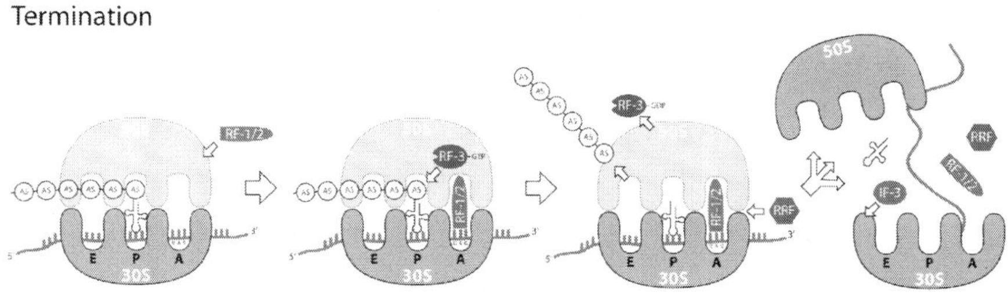

Figure 4.16. Polypeptide Chain Termination

Source: Image by Wikimedia Commons

It would advocate that the polypeptide release specificity mediated by RFs might have evolved autonomously after the phylogenetic domains were separated 2.7 billion years back. Moreover, the fact that the RFs belonging to both the kingdoms would have virtually no sequence or structural homologyemphasizes the view.

Assuming the above case to be a real one, the appearance of prokaryotic RFs and the eukaryotic RFs would be signifying an attractive example of parallel evolution. An interesting question that would come up through such observations would be if the termination signals in the mRNA would be conserved between the two kingdoms.

The results of stop codons would be directly observed by protein factors, instead of a tRNA, as in polypeptide elongation. It would mean that the indication for translation termination might be extended beyond the 3 nucleotides stated in the genetic code.

The pro- and eukaryotes' primary evidence that would have supported the concept has been derived from both bioinformatic and experimental studies.

Bioinformatic analysis of nucleotide frequency related to termination codons in bacteria, would have been majorly based on analyses of *Escherichia coli*, signifiesexistence of bias in the occurrence of particular codon 5′ and 3′ of the stop codon.

A set of studies, in *E.coli*, is considered to have verified experimentally the probability of termination efficiency being correlated with the modifications in the nucleotide sequence 5′ and 3′ of the stop codon. Out of the minimum number of tested sequences, nucleotide sequence 3′ of UAG, UGA and UAA would stop codons.

The result of termination potentiality of the nucleotide exactly 3′ of the stop codon might have been mediated over near cognate tRNA or RF decoding or both. Since it is considered to have been displayed with a zero-length crosslink reagent, hence, the stop codon and 3 nucleotides exactly 3′ of the stop codon would be in closely interacting with the cognate decoding factor in *E.coli* ribosomes termination complexes.

Sequences of 5′ and 3′ concerning UGA stop codons are considered to have shown a cooperative interaction to impact bacterial termination. Additionally, the specification of the last two amino acids pertinent to the nascent polypeptide would have been demonstrated *in vivo* to influence bacterial termination efficiency at UGA and UAG stop codons. The P site tRNA specifications would have also been shown to impact termination efficiency.

Early studies through a small subset of eukaryotic genes would have also exhibited bias in the occurrence of nucleotides 5′ and 3′of stop codons. It resulted in proposing that just like in prokaryotes, the base present after the stop codon was considered to be important for termination efficiency, associating eRF1 to recognize a tetra-nucleotide sequence consisting limited redundancy, and no way among one of the three tri-nucleotide stop codons.

Succeeding studies related to gene sequences in eukaryotes and definite studies in yeast, plants and mammals seemed to have shown a similar bias in nucleotide occurrence in the position after the stop codon.

The decoding release factor in eukaryotes, eRF1, would need a stop codon along with an additional nucleotide to enable termination *in vitro*. Site-directed crosslink studies would show eRF1 being in contact with the triplet stop codon, even though the study is not considered to have investigated the nucleotides following the way it had in the previous equivalent studies in bacteria.

The translation termination performance lies to a limited group of selected eukaryotic sequences is considered to have been examined experimentally in the case of both yeast and mammalian cells. The studies proved that the nucleotide sequences both 5′ and 3′ related to the stop codon would be able to modulate termination efficiency.

As per the bioinformatic and experimental studies related to the nature of the translation termination signal, initiated in a restricted number of both pro- and eukaryotic organisms would recommend that the signal encompasses 5′ and 3′ of the simple triplet codon.

Since a significant number of genome sequences are supposed to have completed, it would be possible to commence a comprehensive comparison of translational termination perspectives both within and between organisms associated with the same or different kingdom.

4.4.6. Prokaryotic translation termination assay

The contexts of termination were examined in an assay that would utilize the IgG -binding B domain belonging from *Staphylococcus aureus* protein A in the form of a reporter gene. *E.coli* cultures were supposed to have been developed to mid-log phase with 0.5 mM IPTG. The reporter gene products (2A′ and/or 3A′), consist of two or three domains, were supposed to have been purified by affinity chromatography on IgG Sepharose, which would be trailed by separation through SDS-PAGE.

The termination efficiency had been determined with the help of the ratio of protein emerging from termination at the first stop codon (2A′) in comparison to total protein revealed from the reporter gene (2A′ + 3A′). the ratio turned out to be separate from the level of expression and optical destiny related to the cell culture of the bacteria. E.coli turned out to be together transformed, while suitable with plasmids which had wild-type RF2, RF2 T246S and RF1.

4.4.7. Eukaryotic translation termination assays

- ***In vitro* Translation**

Large Scale RNA Production Systems, the RiboMAX™ - to synthesis mRNA for translation in Rabbit Reticulocyte Lysate (Promega), T7 (Promega) had been used. Through restriction endonuclease digestion, the dual- luciferase reporter template DNA was made. Moreover, it was transcribed as per the instructions of the manufacturer. Traditionally, 250-

500 ng of RNA was supposed to have been incubated with nuclease-treated rabbit reticulocyte lysate added with amino acids at 30°C for 90 min. translation reactions had been added while suitable with eRF1 purified as per it was described.

- ***In vivo* translation**

According to the instructions of the manufacturer, COS-7 cells had been rapidly transfected through a dual -luciferase reporter plasmid by using FuGENE 6™ reagent (Roche). Samples relating to 2×10^4would have been transfected through 500-1000 ng of plasmid DNA. Cells seemed to be gathered 48 hours after transfection and lysed in 1× Passive Lysis buffer (Gibco).

- Dual luciferase assay

The *Renillareniformis* luciferase (RLuc) 5′ from the trial sequence is supposed to have been assayed through the conditions explained by Matthews et al., 1977 accompanied by the modifications of Srikantha et al.1996 and the *Photinus pyralisluciferase* (Luc^+). Around 3′of the test sequence would have been assayed using the conditions explained by Tanguay and Gallie (1996). The efficiencies of termination are supposed to have been measured from the ratio of $Luc^+$to RLucactivity under the test models being relative to the ratio of $Luc^+$to RLuc within control samples in which the two reporters appeared to be in the same ORF.

4.5. CONCLUSION

The polypeptide chain is synthesized catalytically. The above stated 'factors' would be proteins that would be required for the initiation, elongation and termination of the polypeptide chain. Each of such processes would include one to three factors. There would be a fundamental difference between the eukaryotic and prokaryotic factors.

Protein synthesis would be considered to be a relatively energy-consuming process. Moreover, ATP andGTP would be required. A major section of the energy would be used to analyze. No sooner the peptide is synthesized, it would start folding even before being ready. A secondary and tertiary structure would be developed, achieving, therefore, a thermodynamically favorable state.

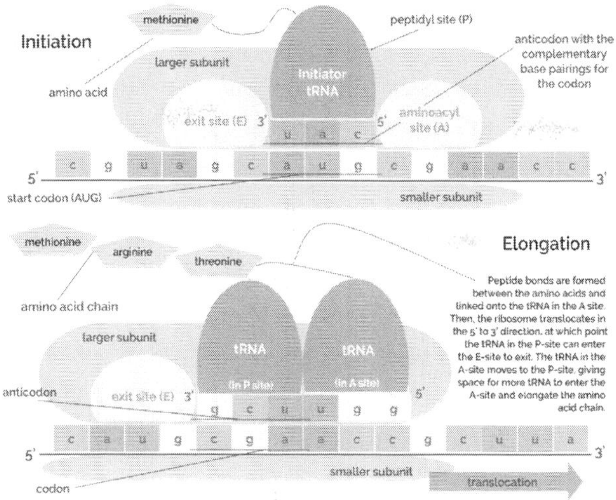

Figure 4.17. Translation – Initiation & Elongation

Source: Image by Wikimedia Commons

The chain would often be modified after the completion of the synthesis. Few parts would be detached, carbohydrates or lipids would be included or the peptide would be activated through acetylation, methylation or phosphorylation. A portion of the nascent protein would stay inside the cytosol, whereas, others would be exported from the membrane or imported into mitochondria or chloroplasts. All such modifications would be termed as post-translational modifications or modulations.

A particular nucleotide sequence would be required for the binding of an mRNA to a ribosome and it would be leading the necessary coding sequence. Due to various recognition signals, factors and other regulating systems, the translation of a eukaryotic mRNA with prokaryotic ribosomes would be unviable (neither would it work in the other way).

All such systems (inclusive of post-translational modifications) collectively would be termed as gene expression. The multiple steps needed and the huge amount of energy consumed would reveal the importance of the process for the primary metabolism of the cell. , Almost 88% of *Escherichia coli* cells' total energy turnover would be required protein biosynthesis.

Knowing the products (proteins, enzymes, regulators) major functions for cell metabolism, its expenses could be fathomed; the convenient operation of the single metabolic routes would be reliant on their function. However, the number of errors would be significantly higher than those of imitation. A host of mRNA and protein replicas per gene would be formed during transcription. Every cell would be able to survive even if a few of them are faulty. However, during replication, the only molecule would be doubled. The various steps of gene expression would allow regulations at various stages. Differential gene expression would make way for achieving differentiation. It implies that the indications contained in the DNA or the corresponding mRNA (nucleotide sequence, secondary structures) would help to determine if a gene would be expressed more frequently than the other or that particular set of genes would be expressed earlier than the other at the time of ontogenesis. There might be abrupt or fluctuating transformations.

The mRNA bonded to the ribosome, for example, would be dependent on its recognition order. However, not every mRNA types would be connected with the same sequence. Consequently, the ones that possess the highest resemblance of the recognition sequence, would mostly bind.

With low resemblance probability, the mRNA would be bonded by a ribosome low, and due to which only some of the protein molecules would be produced. However, it might be one of the various other possibilities available for determining the protein molecules for each cell. An additional control system would be, for example, the resemblance between mRNA and initiation factors, a third would be the half-life of the mRNA.The translation would be similar to transcription and would function according to a specific pattern. Immediately after being detached by the ribosome, mRNA would be bound by an additional ribosome. It would facilitate the synthesis of a second polypeptide chain which would already start before the completion of the first one. Hence, it would help to produce multiple polypeptide chains simultaneously, through the same mRNA template. Several ribosomes would be involved in the process and such complexes would be termed as polysomes.Once the polypeptides would have all the required amino acids, it does not imply that the polypeptide would be completely prepared to carry out its function in the cell. They would often require certain modifications. Amino acids might be chemically changed or removed during and after translation. The new polypeptide would even fold into a specific 3D structure and might bond with other polypeptides forming a multi-part protein.Several proteins would be capable of folding independently, but for some, assistance would be required to prevent them from sticking to one another inaccurately during the complex folding process. Moreover, certain proteins would even contain special amino acid sequences that would guide them to specific parts of the cell. Such a sequence would be mostly located near to the N- or C-terminus.

REFERENCES

1. Berg, J., Tymoczko, J. and Stryer, L., 2002. *Primary Structure: Amino Acids Are Linked By Peptide Bonds To Form Polypeptide Chains.* [online] Ncbi.nlm.nih.gov. Available at: <https://www.ncbi.nlm.nih.gov/books/NBK22364/> [Accessed 4 May 2020].
2. Choi, S., Olsen, D., Roll-Mecak, A., Martung, A., Remo, K., Burley, S., Hinnebusch, A. and Dever, T., 2000. Physical and Functional Interaction between the Eukaryotic Orthologs of Prokaryotic Translation Initiation Factors IF1 and IF2. *Molecular and Cellular Biology*, [online] 20(19), pp.7183-7191. Available at: <https://mcb.asm.org/content/20/19/7183> [Accessed 4 May 2020].
3. Courses.lumenlearning.com. n.d. *Steps Of Translation | Biology For Majors I.* [online] Available at: <https://courses.lumenlearning.com/wm-biology1/chapter/reading-steps-of-translation/> [Accessed 4 May 2020].
4. Easybiologyclass.com. n.d. *Prokaryotic Trasnlation Vs Eukaryotic Translation | Easy Biology Class.* [online] Available at: <https://www.easybiologyclass.com/difference-between-prokaryotic-and-eukaryotic-translation/> [Accessed 4 May 2020].
5. Frank, J., Gao, H., Sengupta, J., Gao, N. and Taylor, D., 2007. The process of mRNA-tRNA translocation. *Proceedings of the National Academy of Sciences*, [online] 104(50), pp.19671-19678. Available at: <https://www.pnas.org/content/104/50/19671#ref-24> [Accessed 4 May 2020].
6. Greganova, E., Altmann, M. and Bütikofer, P., 2011. Unique modifications of translation elongation factors. *FEBS Journal*, [online] 278(15), pp.2613-2624. Available at: <https://febs.onlinelibrary.wiley.com/doi/full/10.1111/j.1742-4658.2011.08199.x> [Accessed 4 May 2020].
7. Hughes, D., 2013. Elongation Factors. *Brenner's Encyclopedia of Genetics*, [online] pp.466-468. Available at: <https://www.researchgate.net/publication/323813013_Elongation_Factors_Translation> [Accessed 4 May 2020].
8. Kaiser, G., n.d. *19.1: Polypeptides And Proteins - Biology Libretexts.* [online] Bio.libretexts.org. Available at: <https://bio.libretexts.org/Bookshelves/Microbiology/Book%3A_Microbiology_(Kaiser)/Unit_7%3A_Microbial_Genetics_and_Microbial_

Metabolism/19%3A_Review_of_Molecular_Genetics/19.1%3A_Polypeptides_and_Proteins> [Accessed 4 May 2020].

9. Nature.com. n.d. *Translation / RNA Translation | Learn Science At Scitable*. [online] Available at: <https://www.nature.com/scitable/definition/translation-rna-translation-173/> [Accessed 4 May 2020].
10. Study.com. n.d. *Polypeptide Chain: Definition, Structure & Synthesis*. [online] Available at: <https://study.com/academy/lesson/polypeptide-chain-definition-structure-synthesis.html> [Accessed 4 May 2020].
11. TeachMePhysiology. n.d. *Translation Of DNA - Initiation - Elongation - Termination - Teachmephysiology*. [online] Available at: <https://teachmephysiology.com/basics/protein-synthesis/dna-translation/> [Accessed 4 May 2020].
12. Themedicalbiochemistrypage.org. n.d. *Protein Synthesis -Translation And Regulation*. [online] Available at: <https://themedicalbiochemistrypage.org/protein-synthesis.php> [Accessed 4 May 2020].
13. Www1.biologie.uni-hamburg.de. n.d. *Botany Online: Molecular Genetics - Protein Biosynthesis - Translation*. [online] Available at: <http://www1.biologie.uni-hamburg.de/b-online/e21/21d.htm> [Accessed 4 May 2020].
14. Brock et al. 1998
15. Barta A, Dorner S, Polacek N, 2001
16. Ogle JM , Brodersen DE , Clemons WM, Jr , Tarry MJ , Carter AP , Ramakrishnan V, 2001
17. Moazed D, Noller HF, 1989
18. Rodnina MV, Savelsbergh A, Katunin VI, Wintermeyer W, 1997
19. Agrawal RK, Penczek P, Grassucci RA, Frank J, 1998
20. Agrawal RK, Heagle AB, Penczek P, Grassucci RA, Frank J, 1999
21. Stark H, Rodnina MV, Wieden HJ, van Heel M, Wintermeyer W, 2000
22. Valle M, Zavialov A, Sengupta J , Rawat U , Ehrenberg M , Frank J, 2003
23. Gao H , Zhou Z , Rawat U , Huang C , Bouakaz L , Wang C , Cheng Z , Liu Y , Zavialov A , Gursky R , et al. 2007
24. Taylor DJ , Nilsson J , Merrill AR , Andersen GR , Nissen P , Frank J, 2007
25. Agrawal RK , Penczek P , Grassucci RA , Burkhardt N , Nierhaus KH

, Frank J, 1999

26. Spahn CM, Gomez-Lorenzo MG, Grassucci RA, Jorgensen R, Andersen GR, Beckmann R , Penczek PA , Ballesta JP, Frank J, 2004
27. Walsh CT, Garneau-Tsodikova S & Gatto GJ Jr, 2005
28. Ahrne E, Muller M & Lisacek F, 2010
29. Yang XJ, 2005
30. Saini P, Eyler DE, Green R & Dever TE, 2009
31. Blaha G, Stanley RE & Steitz TA, 2009
32. Hartman H & Smith TF, 2010
33. Nakamura Y & Ito Km, 2003
34. Yanagisawa T, Sumida T, Ishii R, Takemoto C & Yokoyama S, 2010
35. Matthews et al. 1977
36. Srikantha et al.1996
37. Tanguay and Gallie, 1996

5

Messenger RNA (mRNA) and its Translation

CONTENTS

This chapter is about the translation process that occurs with the help of mRNA. In this chapter, there is a deep discussion about all the procedures that are important for the translation process.

In this chapter, there is an explanation about the mRNA processes used for translation by using initiation, elongation, and termination steps. All the three steps are being explained in a detail. Before starting these steps, it is also important to understand the mRNA and its presence in prokaryotes and eukaryotes.

After that, the process of translation and then the steps initiation, elongation, and termination in detail are explained below.

5.1. INTRODUCTION

From the last two decades ,mRNA is believed to gain growing attention in medical research. It is particularly important as a vehicle that is used to deliver the tumor antigen as well as the activation stimuli that is used for the induction of immune that responses against cancer.

The growing interest in the mRNA is considered as an active pharmaceutical ingredient that is used in anti-cancer immunotherapy and it can be explained by its versatility and by the advantages that it offers. mRNA is considered safe when it is being compared with the viral vector or plasmid DNA.

mRNA is considered to be more important as it lacks the genomic integration capacity and it results only in the transient expression of the encoded protein. Likewise, many cell types can be modified with the mRNA, these cells include those cells which are hard to modify cells, like dendritic cells (DCs).

Plant growth and function are considered that they require highly regulated spatial and temporal regulation of gene expression. The key step that helps in the regulator continuum from the gene to protein to phenotype is the decoding of the mRNA into the polypeptide chain or protein with the help of ribosomes.

The process of translation often requires many of the RNAs, the messenger RNA (mRNA) transcript, transfer RNAs (tRNAs) and the ribosomal RNAs (rRNAs) of the ribosomes. The process of translation also works as a score protein factors that functions as individual protein or either as a multi-subunit complex.

The decoding process of mRNA into a polymer of amino acids is considered as a very ancient process. Such as, the machine that is used for this process, the two subunits ribosomes, is said to conserve across all the forms of life on the Earth.

The three basic phases of the translation process are – initiation, elongation and termination, that are generally conserved. There are many other components of the complex apparatus that are found to be involved are recognizable across the phyla, especially across the animal kingdom, plants and fungal.

Instead of having a conversation about the basic chemistry and protein synthesis, nature has been found to evolve many different ways of starting the first phase which is known initiation. The second phase is known as elongation, in which there is an addition of amino acid to the polypeptide.

The third phase is known as the termination, in this process, there is the completion of the process which is found to be more preserved across all the kingdoms.

Eubacteria have three proteins those are known as initiation factors that helps to unite the mRNA, the initiator tRNA which is usually known as $tRNA_i^{Met}$ and a small subunit of a ribosome. They assemble as a large subunit of the ribosome that starts the process of elongation.

Archaea and eukaryotes are considered to have expanded this machinery to include 10 or more protein complexes. Even though, it is found that *Archaea* lacks the eIF3 and eIF4 families that is present in the eukaryotes.

In addition to this, there are a number of "flourishes" those are being added to the nuclear-encoded eukaryotic mRNA. These features are being added in the nucleus during the process of transcription and are considered to play important role in transcript stability during the journey from the nucleus to the cytoplasm and from the lifetime of the mRNA.

The translation apparatus was found to be involved to use these added mRNA features in order to facilitate the process of initiation in the cytoplasm. In the eukaryotes, the role of the extended cohort of initiation factors is to participate in the exquisitely with the complicated schemes to regulate the process.

What can be more important for a cell than the process of synthesis of proteins that helps in catalyzing the chemistry of the metabolism for making the energy required for the process of cell growth, cell functions and cell division?

It is not surprising that the plant regulation and transitional apparatus is varying from other eukaryotic organisms. This variation occurs due to the specialized cellular biochemistry, environmental plasticity and the developmental complexity that confers the survival and reproduction that is centered around the capture of light energy and its conversion to chemical energy.

5.2. MESSENGER RNA (MRNA)

Messenger RNA is considered in a large family of RNA molecules that are found to be complementary to the DNA molecule. It also conveys genetic information from DNA which is to be translated by the ribosomes into the proteins (Brenner et al., 1961).

mRNAs are like DNA; they are the nucleic acids that contain a specific sequence of nucleotides. These nucleotides are being ‘read’ or translated by the ribosomes to assemble the polymers of amino acids, a protein.

The mRNA is said to play a key role in the ‘central dogma’ of molecular biology by dealing with the transfer of a sequence information from the DNA to RNA to Protein form (Brenner et al., 1961). There are a series of transfer which is possible by this dogma.

The transfer of information from the DNA to RNA via transcription, DNA can be even copied to DNA making DNA replication or protein can be even synthesized from the information in mRNA through translation or a protein synthesis process.

The information rarely gets transferred from RNA to DNA, this usually happens in the case of retroviruses (Baltimore, 1970; Temin and Mizutani, 1970). It is considered that the particles of these viruses like HIV are containing two copies of RNA as their genetic information.

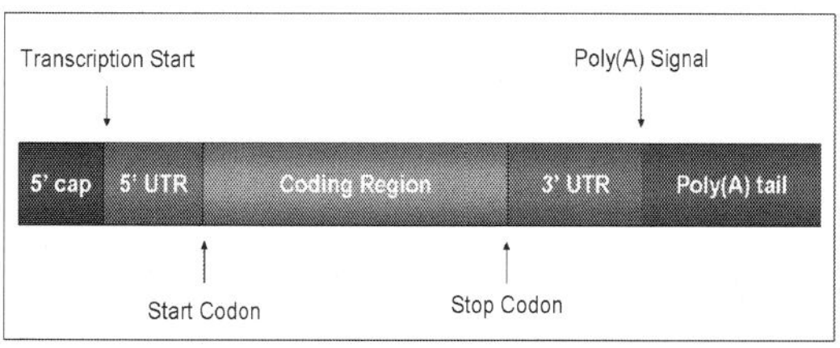

Figure 5.1. mRNA Structure

Source: image by Wikimedia Commons

There are very rare enzymes like reverse transcriptase, who are copying the RNA to a single strand of DNA, which is then integrated into the host's DNA. It is considered that the information is not transferred from the protein to either DNA or RNA.

These transfers often describe that the general flow of the biological information takes place from DNA to RNA to Protein. Messenger RNA is sometimes called as the Rosetta Stone of the biological information. This is called so because it is found to link the transferring genetic information from the DNA into the protein.

While there is a difference in both the structure and mechanism of the function of the prokaryotes and eukaryotic mRNA (Berg, 2002).In the mRNA, the genetic information is being encoded in a four-base alphabet of nucleotides. This forms a codon consisting of three bases each.

Each of the codon's codes for a specific amino acid, with the exception including the stop codons, which will help in determining where the protein synthesis should terminate. The mRNA is being translated by the ribosomes which help in reading the codons.

The start or initiator codon for both the eukaryotes and the prokaryotes is an AUG sequence and this sequence is being read from the 5' to 3' direction. Eukaryotes are generally seen to code for a single protein (monocistronic) (Gerlinger et al., 1977). Whereas, the prokaryotes mRNA is often seen to code for the series of related proteins (Polycistronic) on the same mRNA.

The polycistronic (Gerlinger et al., 1977) mRNA often directs more or less simultaneous synthesis that takes place in each of the encoded polypeptides. For example, trp operon (Hiraga and Yanofsky, 1972) is known as a DNA which is transcribed into mRNA that often codes for the six polypeptides that basically catalyzes the tryptophan synthesis.

Messenger RNA is considered as a short – lived compared to DNA and to the lifetime of the RNA that serves as a regulatory site (Guhaniyogi and Brewer, 2001) of gene expression. After the process of transcription, mRNA molecule can also be edited, processes, and transported before the translation takes place.

Eukaryotic mRNA is considered to require the transport and often seems to be extensive in processing, while the prokaryotes mRNA molecules do not (Clark and Pazdernik, 2013).

5.3. PROKARYOTIC MRNA STRUCTURE

Prokaryotic mRNA is seen to be synthesized in the cytoplasm and they do not require from the nucleus (Clark and Pazdernik, 2013). It is also considered that they do not require the processing and even can begin the translation immediately after the process of transcription gets completed.

Most of the mRNA are containing a sequence at the 5' end of the mRNA which is present at upstream to the AUG start codon, and termed as 5' untranslated region (5' UTR) (Meijer and Thomas, 2002), and the region which is downstream to the stop codon is called as the 3' UTR.

Most prokaryotic mRNA is said to contain a sequence that is in the 5' UTR to position ribosomes for the translation process. This sequence is often named after its discoverers as the Shine – Dalgarno sequence.

5.3.1. Polycistronic mRNA

In the prokaryotic cells, there is a single mRNA that can code for several proteins. Each message present on the mRNA is said to contain in a single 'open reading frame'; whereas, a sequence of codons is seen to be bounded by the start and stop codons.

As there is no such start or stop codon present within the reading frame. The arrangement of the message is considered to be tandem along with a single strand of mRNA that allows the proteins which are also called as gene products is to be translated simultaneously. These gene products are also related to the function.

As mRNA is a single stranded, there are some mRNA molecules that are able to basepair within themselves, they can even form secondary and tertiary three dimensional structures. These structures can also be regulated by the synthesis of polypeptides in the polycistronic mRNA.

One of the examples of this mechanism is the MS2 bacteriophage (Kozak, 1983). The A protein is being coded at the 5' end of the polycistronic message, but it is considered to be needed in only small quantities. The 5' end of the mRNA often seems to be blocked by the tertiary folding of the mRNA that is allowing only limited translations.

This translation is taking place in A protein while allowing the translation that can occur at a more accessible site downstream from the A gene.

5.4. EUKARYOTIC MRNA

In eukaryotes, the things are considered to be little more complex. The molecule that is directly made by the transcription process is called pre-mRNA. That reflects that there are still more steps that are needed to become an actual messenger RNA (mRNA).

These are:

- Addition of a 5' cap at the beginning of the RNA
- Addition of poly-A tail at the end of RNA
- Cutting of introns or junk sequences and pasting together the exons

Once these steps are completed, the RNA turns into a mature mRNA. Further, it can travel out of the nucleus and is used for making it into a protein.

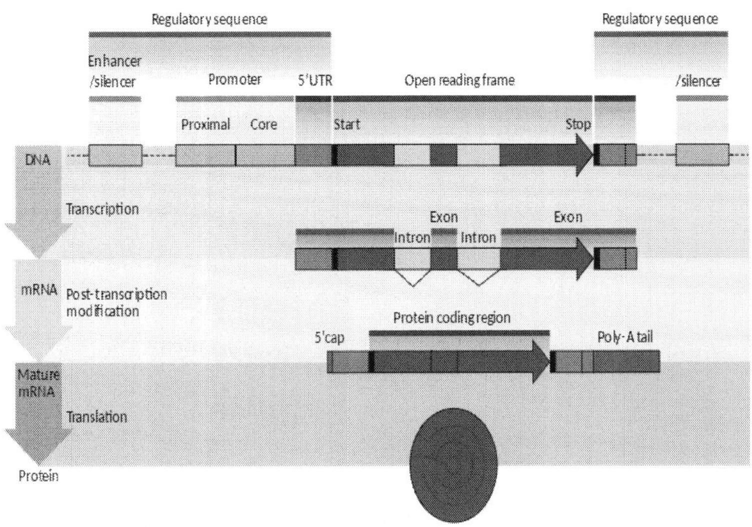

Figure 5.2. Eukaryotic mRNA

Source: Image by Wikimedia Commons

5.4.1. 5' Cap and Poly-A Tail

Both the ends of a pre-mRNA are being modified by the addition of chemical groups. The group that is at the beginning (i.e., 5' end) is called as a cap. Whereas, the group that is present at the end (i.e., 3' end) is called a tail.

Both the cap and the tail are protecting the transcript and it also helps it to get exported from the nucleus and gets translated on the ribosomes which are called as a protein making machine and it is being found in the cytosol.

The 5' cap is being added to the first nucleotide during the transcription in the transcript. The cap is a further being modified guanine nucleotide which helps in protecting the transcript from being broken down. It is also helping the ribosomes to get attach to the mRNA and also starts reading it in order to make a protein.

The above image contains a pre-mRNA with the 5' cap and 3' poly A tail. The 5' cap is present on the 5' end of the mRNA and is considered as a modified G nucleotide. The poly A tail is present on the 3' end of the pre-mRNA and is said to consist of a long string of nucleotides.

5.4.2. RNA Splicing

The third big RNA process event that occurs inside the body is RNA splicing. In this process, specific parts of the pre-mRNA which are also known as introns are being recognized and are removed by a protein and RNA complex, known as spliceosome.

The introns can also be viewed as junk sequences that should be cut out, to have a good parts vision of the RNA molecule that can be assembled further. Whereas there is a piece of the RNA, that is not chopped out are known as exons. These exons are pasted together by the spliceosome to make the final mature mRNA.

The diagram above is showing the exons and the introns. Along with the length of mRNA, there is also an alternating pattern of exons and introns. This pattern is like: Exon 1 - Intron 1 - Exon 2 - Intron 2 - Exon 3.

Each of these is consisting of a stretch of RNA nucleotides. During the process of slicing the introns are being removed from the pre-mRNA. The exons got stuck together to form a mature mRNA that does not contain intron sequences.

5.4.3. Alternative Splicing

Splicing is the process that does allow the process called alternative splicing. In alternative splicing, there is more than one mRNA is being made from the same gene. By using the alternative splicing method, there is a possibility to encode more different proteins that the genes that are present in the DNA.

5.5. IMPORTANCE OF MRNA TRANSLATION

Controlling translation is considered to offer the advantages that take place over the levels of control of gene expression under some particular conditions.

These conditions have been mentioned below.

5.5.1. Rapidity of Response

The rapidity of response is where the cells are needed to increase rapidly the synthesis of proteins, their ability to up-regulate the process of translation of pre-existing mRNAs which are allowing the cell to start to make the corresponding protein. This protein is made without the requirement to activate the transcription, process of the resulting transcript and with that transporting it further to the cytoplasm.

These situations often include the stimulation which takes place by the mitogens, hormones and by the growth factors. They also respond towards the nutrient availability and about the stressful conditions that take place at certain extend.

5.5.2. Independence from Transcription

Where there is little or no ongoing transcription, there the ability to control the process of translation is essential. For example, in the early development of many of the organisms, where it is considered that the transcription is essentially switched off, and in the cell, it has no functional nucleus.

5.5.3. Localized Protein Expression

There are various ways through which the requirement of the protein to the specific locations in the cell is made possible. Firstly, the mRNA is being localized within the cell. Secondly, the protein is made throughout the cell and then it is being transported to its destination.

Thirdly, it localizes the translation process of the mRNA, which may be present widely by itself in the cell. The important examples that can be used for the first and third option are considered to play a key role in controlling the gene expression during the development.

5.5.4. Protein Synthesis as an Anabolic Process

The process of activation of the protein synthesis is also considered as an important response to stimuli which helps in regulating the cell growth. For example, insulin often promotes the process of protein synthesis and anabolic processes like glycogen storage in a variety of cell types and tissues.

5.6. TRANSLATION

The translation is a process which is initiated just after the transcription in the prokaryotes because they do not require the processing or transporting. Therefore, the prokaryotic translation is being coupled to the transcription and occurs co-transcriptionally.

Eukaryotic mRNA is first processed and is being transported to the cytoplasm which is later being translated by either free floating ribosomes or by those attached to the endoplasmic reticulum by the signal recognition particle (Dever and Green, 2012; Hinnebusch et al. 2016; Kozak 1983).

5.6.1. The Genetic Code

It is considered that in an mRNA, the instruction that is used for building a polypeptide often comes in a group of three nucleotides which are known as codons. There are some key features of codons that are described below:

- One codon, AUG, is a "start" signal to kick off translation (it also specifies the amino acid methionine)
- Three "stop" codons mark the polypeptide as finished
- There are 61 different codons for amino acids

These relationships that take place between the mRNA codons and amino acids are known as genetic code.

5.6.2. Codons to Amino Acids

In the process of translation, the codon of an mRNA is being read in order from the 5' end to the 3'end by the molecules which are called a transfer RNA or tRNAs. Each of these tRNA has an anticodon. These anticodons have set of three nucleotides that bind to a matching mRNA codon through base pairing.

The other end of the tRNA is considered to carry the amino acids that are specified by the codon. tRNA often binds with the mRNAs inside of a protein and RNA structure which is called a ribosome. As the tRNA gets to enter in the slots in the ribosome and often binds with the codon.

Their amino acids are linked to a growing polypeptide chain in a chemical reaction. As the end result, polypeptide whose amino acid sequence mirrors with the sequence of the codon in the mRNA.

5.7. TRANSLATION: BEGINNING, MIDDLE, AND END

Translation process occurs in three major stages, which are named as initiation, elongation and termination.

- **Initiation** ("beginning"): initiation is the stage in which the ribosome gets together with the mRNA and the first tRNA, so that translation can begin.
- **Elongation** ("middle"): At the elongation stage the amino acids are brought to the ribosome by tRNAs and linked together to form a chain.
- **Termination** ("end"): In the termination stage, the finished polypeptide is being released to go and do its job in the cell.

5.8. INITIATION

In order to start the translation process, few key ingredients are important for the process. These key points are:

- A ribosome (which comes in two pieces, large and small)
- An mRNA with the instructions for the building of protein
- An "initiator" tRNA, which is carrying the first amino acid in the protein. Which is almost always methionine (Met)

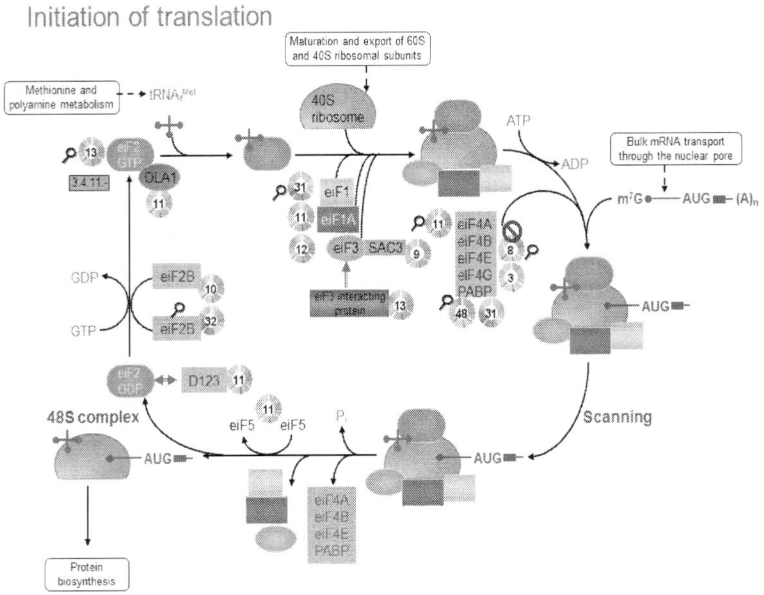

Figure 5.3. Schematic representation of the translation initiation process. *Source*: Image by Material Parasite Metabolic Pathways

5.8.1. Formation of 43S Preinitiation Complexes

Translation initiation often requires a pool of separated ribosomal subunits. The translation is considered as an acyclical process. Ribosomal subunits are found to participate in the initiation which is derived by recycling of the post termination ribosomal complex (post-TCs).

This ribosomal complex comprises 80S ribosomes which are still found to bound to mRNA, P-site deacylated tRNA and with at least one release factor, commonly eukaryotic release factor 1 (eRF1). Post-TCs are being recycled by releasing these dissociating ribosomes and ligands into the subunits.

Recycling is being mediated by eIFs (Pisarev et al., 2007) at the relatively low free Mg^{2+} concentration. eIF3, is consider to dissociate the post-TCs into a free 60s subunits and mRNA and tRNA bound 40S subunits by the cooperation with the loosely associated eIF3j subunit, eIF1 and eIF1A.

Subsequently, it is being observed that eIF1helps by promoting the release of tRNA and eIF3j, which binds with the 40S subunits with the negative cooperativity with the mRNA, which mediates mRNA dissociation.

eIF3 is similar to eIF1 and eIF1A, which remain associated with the recycled 40S subunits, by preventing their re-association with the 60S subunits. Recycling when done at even slightly elevated Mg^{2+} concentration which stabilizes the ribosomal subunit association, will also require ATP binding cassette subfamily E member 1 (ABCE1) .

ABCE1 is considered as an essential ATP binding cassette protein. ABCE splits the post-TC into the free 60S subunits and tRNA and mRNA bound 40S subunits. With them, they are also subsequently releasing the P-site tRNA and mRNA from these 40S subunit which also requires eIF3, eIF1 and eIF1A.

Therefore, eIF3, eIF1 and eIF1A are being recruited to 40S subunits during the process of recycling. Whereas, the eIF2-GTP-Met-$tRNA^{Met}_i$ are subsequently found to be attached to the recycle 40 subunits, with this they also bind simultaneously with the eIF3, eIF1 and eIF1A in order to form the 43S complexes.

There is another protein that can prevent the ribosomal subunits re-associations by simply binding to the 60S subunits is eIF6. But its status which is as an initiation factor is found to be uncertain.

According to some recent studies, it has been yielded insight into the architecture of a 43S complex. It is considered that Eukaryotic and prokaryotic small ribosomal subunits are sharing a common structural core that includes the decoding center.

In this decoding center, the additional eukaryotic ribosomal proteins and 18S rRNA expansion segment are seen to be situated peripherally. 18S rRNA expansion segments are rapidly evolving regions that interspersed throughout the conserved rRNA core that might function in the translation that is eukaryotic specific aspects.

The highest resolution structure of a eukaryotic ribosome is being determined by cryo-electron microscopy (EM). Whereas, the structural homology that is between the eukaryotic and prokaryotic ribosomes are often allowing the high resolution crystal structures of the prokaryotic ribosomes that is to be used to model 40S eIF interactions, based on biochemical data.

The 40S subunit is consisting of a head, body and a platform with the mRNA binding channel which is found to be wrapped around the neck. The bulk of the five- lobed eIF3 molecules are seen to bind with the 40S subunits sides which are found to be facing the solvent; whereas, the eIF3j C-terminal domain localizes in the mRNA binding channel present in the A site area on the inter-subunit side. Even though the eIF2's structure is also available, and the position of the eIF2-GTP-Met-tRNA$^{Met}_{i}$ which is present on the 40S subunits has not been determined. However, in the 43S complexes, the Met-tRNA$^{Met}_{i}$ anticodon loop is being inserted less deeply into the P-site, than in the ribosomal complexes with the established codon-anticodon base pairing.

Its acceptors-end that might be rotated towards the E-site (Allen et al., 2005). eIF1 is found to bind the interface that is between the platform and Met-tRNA$^{Met}_{i}$. the eIF1Ais a structured domain that resides in the A-site, which is forming a bridge over the mRNA Channels.

Whereas, the N and C -terminal tails seem to extend into the P-site. The binding of eIF1 and eIF1A to the 40S subunits seems to induce the conformational changes which are basically involving the opening of the mRNA entry channel 'latch' which is formed by the helix (h) 18 in the body.

Whereas, h34 and rpS3 are found in the neck and is establishing the new head-body connection that is on the solvent side which is basically between the h16 and rpS3.

5.8.2. Attachment of 43S Complexes to mRNA

Though, the 43S complexes are found to be intrinsically capable of 5' -end dependent attachment to the model mRNA with the completely unstructured 5' -UTRs, natural 5' -UTRs. These UTRs are possessing sufficient secondary structure that is loading the 43S complexes onto them; are requiring the cooperative actions of eIF3 and eIF4B or eIF4H.

These actions unwind with the 5' cap proximal region of the mRNA to prepare it for the ribosomal attachments. eIF4F even comprises the cap which is binding the protein eIF4E, the DEAD box RNA helicase eIF4A and eIF4G, are found to function as a "scaffold" that usually binds the eIF4A, eIF4E, the poly-A- binding protein (PABP) and eIF3.

eIF4B and eIF4H are found to enhance the eIF4A's helicase activity that basically contains RRM domains and are also homologous over the entire length of an eIF4H. The cap even gets stacked between the two tryptophan residues on the eIF4E's concave surface.

There are some additional contacts with the cap-proximal nucleotide which is stabilizing the eIF4E's binding of the capped mRNA. A segment of eIF4G is found to get wrapped around the eIF4E's N-terminus which is inducing the structural changes that enhances the eIF4E's cap binding affinity.

eIF4A has two domains and alternates that are between the inactive 'open' and an active 'closed' conformation. In this conformation, both the domains are forming a contiguous RNA binding surface and they also the ATP binding site which is at the domain interface.

It is considered that the low individual helicase activity of eIF4A is quite strongly enhanced by the eIF4G and eIF4B or eIF4H. eIF4G's HEAT – 1 domain often stimulates the eIF4A's helicase activity by simply aligning the DEAD box motifs in both the domains in a productive conformation.

On the other hand, HEAT – 2 also binds with the eIF4A which also plays a modulatory role. The topology of eIF4A – eIF4G – eIF4H complexes indicates that eIf4H binds with the single stranded mRNA behind the eIF4A, which is relative to the direction of the helicase translocation.

This is often suggesting that the eIF4H or eIF4B can also stimulate the eIF4A helicase activity by preventing the mRNA re-annealing and also promoting the progressive unidirectional eIF4A movement. As it was already suggested that because of the high but limited processivity of the eIF4F – eIF4B or eIF4H complexes.

eIF4A is finally seen to dissociate from the mRNA but they are being anchored to its 5' -end by the eIF4E cap interaction. These complexes often resume that there is another cycle of the unwinding which is done by keeping the 5' proximal region constantly to prepare for the ribosomal attachment that is likely to get facilitated by the eIF3 – eIF4G interaction.

Therefore, the recruitment of the 43S complexes is ultimately achieved by the cap – eIF4E – eIF4G – eIF3 – 40S chain of interactions. The 'open – latch' conformation of the 40S subunits I is being induced by the eIF1 and eIF1A is found to be strongly conductive for the attachment.

Despite these advances, the position of the eIF4E is present in the ribosomal complexes and mechanistic aspects about how the mRNA is entering the mRNA binding channels remain unknown. If the cap – eIF4E interactions get persists during the attachment, rarely, the eIF4E bounds with the mRNA can also be threaded through the entire mRNA binding channels.

Therefore, the loading of the 43S complexes is found to be more compatible with the direct positioning of the eIF4E – cap which is at the channel's E – site side. This will often raise the question which is related to from which the nucleotide does 43S complexes which begin inspecting the 5' –UTR during the scanning?The efficiency of the initiation on the mRNAs with the short 5' -UTRs is being invested in the Cell-free system. However, the data is being obtained which often shows that about 50% of the ribosomes bypass an AUG codon which is located within the 12 nucleotides of the cap.

This would reflect that there is the susceptibility of the initiation complexes which will potentially form close to the mRNAs 5' –end to the dissociation by eIF1 rather than the $tRNA^{Met}_{i}$ inability to inspect the mRNA from some certain position.

5.8.3. Ribosome Scanning

After the attachment, the 43S complexes scan the mRNAs downstream of the cap, to the initiation of the codon. Scanning consists of two linked processes: first is the unwinding of the secondary structure in the 5' -UTR and secondly, the ribosomal movement that takes place along it.

43S complexes can also scan the unsaturated 5' -UTRs without the factors which get associated with the RNA unwinding. Thus, they are intrinsically capable of movement along the mRNA. The omission of eIF1A is substantially reducing the ability and lack of eIF1 which almost abrogates this ability. This is also indicating that the movement of 43S complexes also requires for scanning the competent conformations that are induced by the eIF1 and eIF1A.

Although the eIF3 is found to be indispensable for the 48S complex formation, it is even difficult to separate the role in scanning from the functions such as ribosomal recruitment of eIF2 – GTP – Met-$tRNA^{Met}_{i}$ and get attached with the 43S complexes.

However, the eIF3 get interacted with the mRNA upstream of the E-site that is forming an extension of the mRNA-binding channel that might also contribute to scanning, although, the scanning procedure of 5' -UTR which is containing weak secondary structure often requires the ATP and eIF4A, eIF4B and eIF4G.

However, the requirement for the ATP and eIF4A is found to be proportional to the degree of secondary structure. Therefore, in addition to the promoting attachment, eIF4G, eIF4A and eIF4B are found to assist the 43S complexes during scanning.

However, the mechanism uses these factors for assisting the scanning that remains unknown. The cryo-EM-based modelling is being placed by eIF4G at the 40S subunit trailing the edge near the E-site. These are needed to be consistent with the eIF4A,eIF4B, and eIF4G which are acting by the helicase mediated "ratcheting" of mRNA.

This process takes place by the mRNA binding channel, whereas the mRNA secondary structure would be unwound by the 40S subunits. However, there is an alternative model in which eIF4A, eIF4B, and eIF4G unwind the mRNA before it enters the channel.

Although, the ribosomal attachments are being achieved by the cap – eIF4E – eIF4G – eIF3 – 40S chain of the interactions. The fate of each of the link during the transition from the attachment to scanning and during the scanning *per se* is unclear.

It is considered that their maintenance can also cause 5' -UTR towards the 'loop out' which is often allowing only one 43S complex to scan at a time; whereas, the breaking of even one link can also permit the multiple complexes for scanning simultaneously on a single 5' UTR.

There is another important question that concerns with the directionality of the scanning. There is the fact that the initiation frequency is at the 5' proximal AUG. This initiation is significantly reduced by the presence of a nearby downstream AUG.

Initiation frequency often suggests that scanning may also consist of a forward (5' → 3') thrust which is altering with the limited relaxation over a distance of a few nucleotides in the reverse direction.

According to the recent data which is obtained by using the yeast and mammalian systems is suggesting that there is initiation involving the other DEAD box family members. This DEAD box family member in addition to the eIF4A, eIF4A can act with the p97, a eIF4G related protein.

The mammalian DExH-box protein DHX29 gets bind to 40S subunits directly and is often considered to require for the efficient scanning process takes place through the highly structured 5' -UTR *in vitro*. *In vivo*, the silencing DHX29 impaired with the translation which results in the polysome disassembly and accumulation of the mRNA which is free 80S monomers.

DHX29 is being suggested to increase the scanning processivity, which is done by influencing the mRNA binding channel's conformations which is at its entrance. Yeast DEAD-box helicase Ded1 has been implicated in initiation. Where Ded1 is likely seen to be more potent helicase than the eIF4a and their functions are not even redundant.

It is being suggested that the eIF4A helps in promoting the ribosomal attachment, whereas, the Ded1 assist the scanning which is particularly on long 5' -UTRs. The involvement of the initiation of DDX3 and a mammalian Ded1 homologue is seen to be more controversial.

Even though, there is some data which are suggesting that DDX3 depletion is specifically affecting the translation process of mRNAs with the long structured 5'-UTRs. P97, which is ubiquitously expressed in the tissue of the mammals and birds, is found to be homologous towards the C – terminal two-third of the eIF4G.

They even bind the eIF4A and eIF3, but it is considered that they lack an eIF4E binding region. P97 activates the translation of the uncapped mRNA which takes place *in vitro.* Its role during the translation of capped mRNA is not redundant with the eIF4G.

Even though, the depletion of the eIF4G1 and p97 is individually impairing the global translation by 20 to 30 percent approximately; the co-depletion is being reduced by proximately 60%. The depletion of eIF4G1 is selectively impaired with the translation of the mRNAs that are containing the upstream Open Reading Frames (uORFs). This is often suggesting that, these factors are promoting the initiation of the different classes of mRNAs.

5.8.4. Initiation Codon Recognition

In order to ensure the fidelity of the initiation, scanning complexes should have a discriminatory mechanism that will prevent the partial base pairing of the triplets in the 5' -UTR with the Met-$tRNA^{Met}_{i}$ anticodon and it also promotes the recognition of the correct initiation codon.

Which is usually considered as the first AUG triplet in an optimum context GCC(**A/G**)CCAUG**G**, with the purine in the -3 and 'G' in the +4 positions which is relative to the A of the AUG codon simply designated by +1.

eIF1 plays a key role in maintaining the fidelity of the initiation. It also enables the 43S complexes to discriminate against the non-AUG triplet and AUG triplets have a poor context or are found to be located within the 8 nucleotides of the mRNA 5' -end.

eIF1 also dissociates the ribosomal complexes that are aberrantly assembled at such triplets basically in its absence (Pestova TV et al., 2002, Pisarev AV, Kolupaeva VG and co-workers, 2006*).* In the current model, eIF1 with the eIF1A promotes the scanning-competent 'open' conformation of the 43S complex.

But, in order to establish the stable codon – anticodon base pairing, the ribosomal complexes undergo several conformational changes that are antagonized by the eIF1. Establishment of codon-anticodon base pairing is being accompanied by the tightening of the eIF1A – 40S interaction and eIF1's displacement from near the P-site.

These switch the complexes to a 'closed' conformation which is locked onto the mRNA. It is being noticed that yeast eIF1 mutant often dissociates more rapidly from the 48S subunits are enhancing the initiation at the non-AUG codons.

The eIF1A's N-terminal tail (NTT) and the C-terminal tail (CTT), which reaches into the P-site and is considered to have opposite effects on the start-codon selection. These effects are: the CTT often increases its stringency and is even supposed to promote the "open" conformation of the scanning complexes. Whereas, the NTT decreases the accuracy of the initiation and it also promotes the "closed" conformation.

The purines which are -3 and +4 positions are likely to get affected by the initiation codon selection that takes place by stabilizing the conformational changes that occurs upon the codon-anticodon base pairing. This happens by interacting with the eIF2βsubunit of the eIF2 and $AA_{1818-1819}$ in the helix 44 of the 18S rRNA.

In the absence of eIF1's, the stability of the 48S complexes is not being challenged. The complexes with the partial base pairing can be even formed and can participate in the subsequent steps in the translation. So, it is being considered that such complexes cannot maintain their conformation upon the binding of the eIF1, and then the mis paired tRNA is being ejected.

5.8.5. Commitment of Ribosomes to a Start Codon

The initiation of the codon recognition is being followed up by the step during which the arrested ribosomes may also become committed towards the initiation at that specific codon. The commitment step is being mediated by the eIF5, and an eIF2 specific GTPase activating protein (GAP).

eIF5 binds with the eIF2's b-subunit, but it induces the GTPase activity of the eIF2's γ-subunit, which is only present in the eIF2-GTP-Met-$tRNA^{Met}_{i}$complexes that are found to be bound with the 40S subunits. eIF5 has been proposed to act as a classical GAP by simply providing an arginine finger.

There is an alternative hypothesis which suggests that eIF5 derepresses eIF2γ GTPase activity. Premature hydrolysis of eIF2 often bound to the GTP in the 43S complexes, and particularly with the subsequent P_i release, which is being prevented by the eIF1.

Establishment of the codon-anticodon base pairing results in eIF1 displacement, which in result relieves the repression of the GTP hydrolysis and P_i release. Therefore, in addition to its role in the initiation of the codon selection during the 48S complex formation, eIF1 is also maintaining the initiation of the fidelity at a later stage. This process is done by simply linking the hydrolysis of the eIF2-bound GTP with the establishment of the codon-anticodon base pairing.

5.8.6. Ribosomal Subunit Joining

Joining of the 60S subunit and dissociation of the eIF1A, eIF1, eIF3 and residual of an eIF2-GDP is being mediated by the eIF5B. eIF5B is a ribosome dependent GTPase that is considered to be homologous to the prokaryotic initiation factor IF2.

It is considered that hydrolysis of eIF5B-bound GTP does not require for the subunit joining, but it is essential for the eIF5B own release from the assembled 80S ribosomes. eIF5B occupies the same region in the inter-subunit cleft as an IF2, which was being proposed to promote the subunit joining by the help of burying large solvent accessible surfaces on both the subunits.

It is believed that eIF5B alone can displace the eIF2-GDP partially from the 40S subunits; whereas, the complete dissociation will occur only in the presence of 60S subunits especially during the actual subunit joining the event.

Therefore, the interaction of the eIF5B's CTD with the eIF1A CTT, which is likely said to become possible only after the displacement of the latter from the P-site upon the initiation codon recognition. It is therefore required for the efficient subunit joining and GTP hydrolysis by the eIF5B.

This is indicating that the eIF1A often remain associated with the ribosomal complexes throughout the subunit joining process and dissociation from the assembled ribosomes with the eIF5B.

Those eIFs that are found to bind to the 40S subunit interface must be released before or at the subunit joining. The dissociation of the eIF3 and eIF4G is being discussed below.

5.8.7. Re-Initiation After a Short Upstream ORF

There are about 45 to 50% of the mammalian genes that are encoding mRNAs, are considered to have at least one short uORF (typically <~30 codons) upstream of the main protein-coding ORF. In these cases, there are some of the ribosomes (usually <50%) which have been found who have translated the uORF which may be resumed the scanning and reinitiate at the downstream sites.

The post-termination event that occurs at uORF stop codons is probably found to proceed conventionally with the release of the 60S subunit, which is being followed by the deacylated tRNA. But some 40S subunits remain on the mRNA and further resume scanning.

At this stage, there are such 40S subunits that are incompetent for the re-initiation because they are found to lack an eIF2-TC. But this process does not prevent them from scanning during which there is new eIF2-TC can be acquired.

eIF2-TC availability is being determine that how far the 40S subunits migrate before acquiring the one. Rescanning and re-initiation efficiency decreases abruptly with the increasing length of the uORF, if it includes the stable RNA secondary structures that are causing the pausing of the elongation.

This often suggested that it is the time that is taken to translate the uORF that is found a bit critical, rather than the length *per se*. This leads towards the idea about the thought that rescanning depends on some of the eIF ribosome interactions that are promoting the initiation at the uORF AUG, it was the persisting for the time that is taken to complete the uORF translation.

The indication is the critical interactions of those involving eIF4G, because of the re-initiation which is observed only if eIF4F and eIF4B or even at a minimum eIF4G p50 fragment with the eIF4A and eIF4B which are participating in the primary initiation process or event which occurs at the uORF AUG.

As, eIF3 binds mainly with the 40S subunit solvent face, which by no means of the eIF3 - 40S contacts which will allow the subunit joining.

eIF3 can be therefore remain bounded transiently with the 40S subunit in a metastable state. The eIF4G – eIF3 interactions were still considered to be in place by the time uORF translation that had been completed. It can also retain the post-termination 40S subunit which is on the mRNA and even promotes its rescanning.

5.9. ELONGATION

Translation elongation is considered as an evolutionarily conserved progression of the ribosome that is being catalyzed by the polypeptide formation which takes place through mRNA decoding. After completion of the subunit joining with the Met-$tRNA^{Met}_i$ in the P-site of the ribosome, then the second codon which is in the A-site await for the interaction that takes place with the anticodon of an aminoacyl-tRNA (aa-tRNA) which is then coupled to the eEF1A GTP (Dever and Green, 2012).

The presence of appropriate codon-anticodon interactions occurs at the A-site will often stimulate the peptidyl transferase reaction that generates a peptide bond that takes place between the Met-$tRNA^{Met}_i$ and the aa-tRNA. They will leave a deacylated $tRNA^{Met}_i$ in the P-site.

Subsequently, the translocation of the mRNA by one codon will shift the peptidyl tRNA to the P-site and another codon with the deacylated tRNA to the E-site, which will help in the A-site for the next appropriate aa-tRNA and continuation of the cycle (Dever and Green, 2012; Doerfel et al., 2013). Translocation is being facilitated by the eEF2-GTP binding and GTP hydrolysis.

The principal that is used for casting this process often includes the aa-tRNAs, eEF1B (homolog of bacterial EF-Ts), the ribosome, eukaryotic elongation factor (eEF)1A (homolog of bacterial EF-Tu), and eEF2 (homolog of bacterial EF-G).

The role played by eEF5 in the process of elongation is considered as an emerging process and like its prokaryotic homolog appear in the involvement of elongation of the amino acid sequence which is enriched in runs of proline or glycine (Doerfel et al., 2013; Gutierrez et al., 2013; Ude et al., 2013).

Aminoacyl tRNA synthetase (aa-synthetases) often seems to participate in the backstage. These enzymes are being encoded by a large family of the nuclear gene present in Arabidopsis that couples with the cognate tRNAs to their amino acids to form the acetylated tRNA known as aa-tRNA in an ATP dependent reaction.

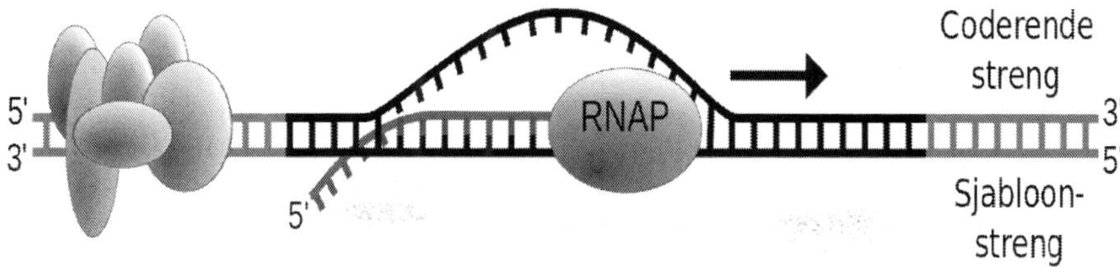

Figure 5.4. Schematic representation of elongation process.

Source: Image by Wikimedia Commons

5.9.1. eEF1A

eEF1A is considered as an ortholog of a bacterial elongation factor – Tu (ED-Tu). This factor is forming a ternary complex with the help of GTP and aa-tRNA, this helps in delivering to the peptidyl transferases center when there is corresponding codon that is present in the A-site.

Initially, loose binding is being followed by a recognition event that involves the hydrolysis of the GTP and structural rearrangements of the tRNA, ribosome and eEF1A. eEF1A-GDP is then released for recycling by the eEF1B.

eEF1A is considered as a highly abundant protein that constitutes about 1% of the total protein in a cell. The four paralogs are encoding the protein present in the Arabidopsis. *Maize opaque 2*seed endosperm mutant has been found to increase the level of the eEF1A from the multiple genes and is being associated with the improved lysine content (Lopez-Valenzuela et al., 2003; Lopez-Valenzuela et al., 2004).

Interestingly it has been found that eEF1A has a function and interactions outside of its role in the translation which includes association with the cytoskeleton, proteolysis, viral propagation, nuclear export and apoptosis (Browning, 1996; Sasikumar et al., 2012).

This factor is also known for participating in the processes including the export of tRNAs from the nucleus and even targeting the damaged and misfolded proteins towards the proteasome (Sasikumar et al., 2012). Whereas EF1A is said to be reported for having the interactions with the tombus-virus replication complex and the 3' tRNA is like a structure of a turnip yellow mosaic virus (Matsuda et al., 2004; Li et al., 2009).

5.9.2. eEF1B

In the prokaryotes, EF-Tu-GDP without the assistance from the EF-Ts, it cannot recycle the GDP to GTP. Similarly, eEF1A is considered to require an exchange factor, eEF1B. In the contrast to a single polypeptide EF-Ts, eEF1B is considered as a complex of protein that varies in complexity from the eukaryote to eukaryote.

eEF1B is having three components in the plants, eEF1Ba, eEF1Bbb and eEF1B. In these components, two of them are found in yeast and one is found in the mammals (Sasikumar et al., 2012). In some of the organisms, the eEF1B also includes a valyl tRNA synthetase.

There are very little who know specifically about the eEF1B from the plants, other than it is being purified from the wheat germ (Lauer et al., 1984). It has also shown that it plays an important role in viral replication (Sasvari et al., 2011; Hwang et al., 2013), that are similar to eEF1A (Matsuda et al., 2004; Li and Nagy, 2011) and the cap binding complex subunits (Wang and Krishnaswamy, 2012).

5.9.3. eEF2

eEF2 is considered as the functional equivalent to the EF-G of the prokaryotes. Peptide bond formation occurs rapidly by following the acceptance of the aa-tRNA into the A-site of the peptidyl transferase center, that is within the large ribosomal subunit.

This region is being largely comprised of rRNA and is considered to be highly conserved between the prokaryotic and eukaryotic ribosomes. This is simply indicating the process of peptide bond formation that is quite ancient (Dever and Green, 2012).

After the peptide bond formation that occurs in the peptidyl transferase reaction which is catalyzed by the ribosome. It is also necessary to move the uncharged tRNA from the P-site into the E-site, to free the A-site for the upcoming aa-tRNA.eEF1A.GTP.

A GTP.eEF2 complex often binds with the ribosome and its GTP hydrolysis which promotes the movement of mRNA.tRNA.tRNA hybrid which is being forward by the three nucleotides which are coinciding with the movement of the P-site deacylated tRNA towards the E-site and with this also ejecting the deacylated tRNA from the E-site.

eEF2 is considered as the post translational modification done at a conserved histidine residue to the diphthamide. This modification makes the eEF2 that is the target for the ADP ribosylation that takes place by diphtheria like toxins (Ortiz et al., 2006; Zhang et al., 2008a).

For this unusual modification, the biological significance is unknown, despite its conservation that takes place across all the eukaryotes and *Archaea*. eEF2 is also known as a substrate that is used for the phosphorylation by using Ca^{2+}/calmodulin-dependent eEF2 kinase (eEF2K). This will reduce the eEF2 association with the ribosome.

The phosphorylation site present in the mammals is found to conserve the threonine residue (T66). The mammalian AMP kinase and mTOR signaling pathways often converge to inhibit the Ca^{2+} dependent eEF2K activity. Therefore, there is limited translation elongation under the nutrient limiting the conditions (Leprivier et al., 2013).

5.9.4. eEF5 (nee eIF5A/eIF4D)

eEF5 was also known as eIF5A/eIF4D, because of its initial report as a stimulator of Met-puromycin synthesis that takes place *in vitro*, which is a model assay for the initiation. eEF5 was recognized later as a facilitator of elongation (Nanda et al., 2009) and was being confirmed that the functional and structural equivalent to the elongation factor P of the eubacteria.

Both the eEF5 and EF-P are seemed to be involved in the efficient elongation of the proteins with the run of glycine or proline residues (Doerfel et al., 2013; Gutierrez et al., 2013; Ude et al., 2013).

5.10. TERMINATION

The process if elongation ends when the translocation takes place in one of the three stop codons, which are UAA, UAG or UGA, into the A-site of the ribosome. This initiates the process of terminations, which ends the disengagement of the peptide from the ribosomes (Dever and Green, 2012; Jackson et al., 2012).

There are two eukaryotic releasing factors known as eRF1 and eRF3 in the plants, the same case is in mammals. The process of the translation complex upon the termination is still considered to be lacking in the molecular details.

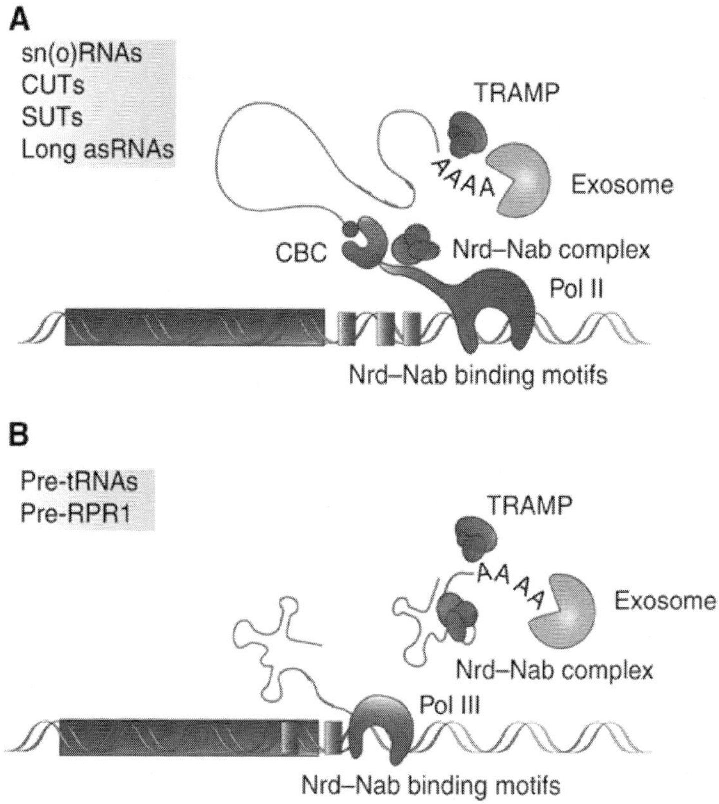

Figure 5.5. Schematic representation of the Termination process.

Source: Image by ResearchGate

The termination process is usually followed by the ribosome release, but it may be also followed by the re-initiation of the translation after a short coding sequence (Roy and von Arnim, 2013). In some of the special case, like premature termination at a nonsense codon, degradation process of the transcript which occurs via NMD pathway that is discussed below.

5.10.1. eRF1 and eRF3

eRF1 is considered to be evolutionarilyy related to the bacteria RF1 and RF2, from the class RFs. The structure of these proteins are often considered to resembles a tRNA that is allowing the N-terminal domain of the RF to dock at the A-site and get directly interacted with the stop codon (Dever and Green, 2012; Jackson et al., 2012).

The high fidelity of this binding is said to get coupled with the GTPase activity of the eRF3. eRF3 which promotes the hydrolysis of the peptidyl tRNA that is necessary for releasing the polypeptide from the P-site and the ribosome (Dever and Green, 2012; Jackson et al., 2012).

At the structural level, it is being found that N-terminal region of the eRF1 binds with the stop codon. After that, the middle domain gets an entry in the peptidyl transferase center where it is found to promote the hydrolytic release of the polypeptide.

On the other hand, the C-terminal region gets interacted with the eRF3. eRF3's GTPase activity that is necessary for both of them to increase the rate of the peptide hydrolysis by the eRF1 and therefore, the efficiency of the termination (Dever and Green, 2012; Jackson et al., 2012).

The structure of the eRF3.GTP.eRF1 is a ternary complex present on the ribosomes which were determined, which is revealing the number of features that are being suggested that it has a very similar GTPase activation mechanism that is used for the prokaryotic aa-tRNA.EF-Tu.GTP complex (des Georges et al., 2014).

eRF1 is being retained after the termination and it is also important for recycling the ribosome by recruiting the ABCE1/RIL1 protein which often functions with the eIF6 in the dissociation of the ribosome into the subunits for recycling (Pisarev et al., 2010).

Yeast eRF1 is considered to have an additional function that may affect the cytoskeleton and the cell cycle (Valouev et al., 2002).

Arabidopsis has three *AteRF1* genes that are encoding the functional RFs (Chapman and Brown, 2004). The over-expression of the *AteRF1-1* often results in the silencing of the *AteRF1-1* and to some extent *AteRF1-2* and *AteRF1-3*.

These effects are causing the phenotype *broomhead,* which means altered spacing that takes in between the inflorescence stem causing a broom like structure. These effects are associated with the perturbations in the cell elongation and cell division (Petsch et al., 2005).

AteRF1-2 mRNA levels are being induced by the high glucose levels and *AteRF1-2* are overexpression of the lines which are displaying the increased glucose-mediated repression of the germination (Zhou et al., 2010b).

These genotypes are also considered as a hypersensitive towards the paclobutrazol, which is an inhibitor of the gibberellin biosynthesis, as well as abscisic acid. t-DNA insertion mutants of *AteRF1-2* have shown the resistance towards the gibberellin synthesis inhibitors during the process of germination.The role played by *AteRF1-2* is still not clear in the glucose sensing or phytohormone responses which reflect its role played in the termination or some other cellular roles. It is important to determine whether the *broomhead* and other phenotypes are associated with the eRF1 mutants in *Arabidopsis* which are related to the translation or other processes.

5.11. CONCLUSION

Protein synthesis is considered as a fundamental process, which is being used in the process of gene expression. The function of gene expression depends upon the amount and the availability of the mRNA transcript as well as on the activity of many of the protein and RNA-protein complexes.

In this chapter, the main focus is on the mechanism of the mRNA translation that is played for expressing the gene. Firstly, it is important to understand about mRNA and role of mRNA in prokaryotes and eukaryotes.

With this, it is also important to understand the importance of mRNA in the process of translation. The process of translation takes place in three parts. These parts are initiation, elongation, and termination. These processes occur in the presence or with the help of ribosome.

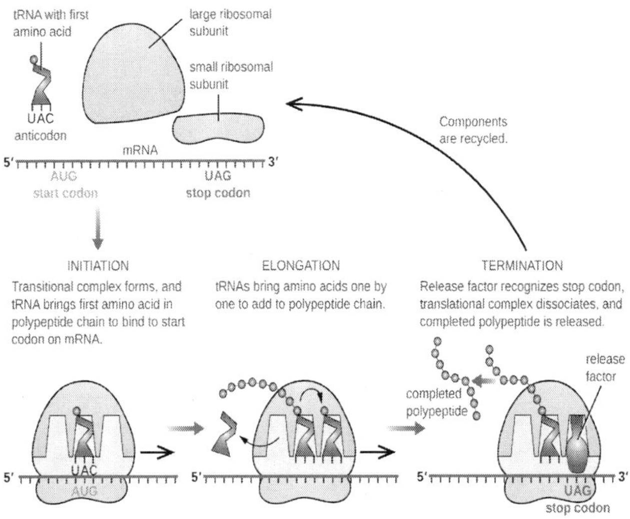

Figure 5.6. Schematic representation of Translation process.

Source: Image by Wikimedia Commons

Different parts are involved in the process of initiation, like the formation of 43S pre-initiation complexes, then attachment of this complex to mRNA and likewise there many steps that help in the process of initiation.

Likewise, the elongation process is also followed by different steps like the role of eEF1A, eEF1B, eEF2, eEF5. These assist in the process of elongation.

Finally, the process of termination is when the stop codons get attached with the A-site of the ribosome, then that initiates the process of termination. Therefore, initiation, elongation and termination are important for the process of translation that takes place through mRNA.

REFERENCES

1. Arayici, P., Acar, T. and Karahan, M., 2014. *Transcription And Translation*. [ebook] Available at: <https://www.researchgate.net/publication/325827703_Transcription_and_translation> [Accessed 8 May 2020].
2. Browning, K. and Bailey-Serres, J., 2015. Mechanism of Cytoplasmic mRNA Translation. *The Arabidopsis Book*, [online] 13, p.e0176. Available at: <https://www.ncbi.nlm.nih.gov/pmc/articles/PMC4441251/> [Accessed 8 May 2020].
3. Domashevskiy, A., 2016. *Messenger RNA (Mrna): The Link Between DNA And Protein*. [ebook] Available at: <https://www.researchgate.net/publication/305698915_Messenger_RNA_mRNA_The_Link_between_DNA_and_Protein> [Accessed 8 May 2020].
4. Jackson, R., Hellen, C. and Pestova, T., 2010. The mechanism of eukaryotic translation initiation and principles of its regulation. *Nature Reviews Molecular Cell Biology*, [online] 11(2), pp.113-127. Available at: <https://www.ncbi.nlm.nih.gov/pmc/articles/PMC4461372/> [Accessed 8 May 2020].
5. Khan Academy. n.d. *Eukaryotic Pre-Mrna Processing | RNA Splicing (Article) | Khan Academy*. [online] Available at: <https://www.khanacademy.org/science/biology/gene-expression-central-dogma/transcription-of-dna-into-rna/a/eukaryotic-pre-mrna-processing> [Accessed 8 May 2020].
6. Khan Academy. n.d. *Stages Of Translation (Article) | Khan Academy*. [online] Available at: <https://www.khanacademy.org/science/biology/gene-expression-central-dogma/translation-polypeptides/a/the-stages-of-translation> [Accessed 8 May 2020].
7. Lodish, H., Berk, A., Zipursky, S., Matsudaira, P., Baltimore, D. and Darnell, J., 2000. *Processing Of Eukaryotic Mrna*. [online] Ncbi.nlm.nih.gov. Available at: <https://www.ncbi.nlm.nih.gov/books/NBK21563/> [Accessed 8 May 2020].
8. Mishra, A., 2017. Messenger RNA. *Encyclopedia of Animal Cognition and Behavior*, [online] pp.1-5. Available at: <https://www.researchgate.net/publication/321025992_Messenger_RNA> [Accessed 8 May 2020].
9. n.d. *DNA Function*. [ebook] Available at: <https://facultystaff.richmond.edu/~lrunyenj/bio554/lectnotes/Chapter13.pdf> [Accessed

8 May 2020].

10. n.d. *DNA, RNA, Replication, Translation, And Transcription*. [ebook] Available at: <https://sites.engineering.ucsb.edu/~shell/che170/DNA-notes.pdf> [Accessed 8 May 2020].
11. Potter, K., Cremona, N. and Wise, J., 2013. Messenger RNA Processing in Eukaryotes. *Encyclopedia of Biological Chemistry*, [online] pp.59-64. Available at: <https://www.researchgate.net/publication/323706474_Messenger_RNA_Processing_in_Eukaryotes> [Accessed 8 May 2020].
12. Proud, C., 2001. Regulation of mRNA translation. *Essays in Biochemistry*, [online] 37, pp.97-108. Available at: <https://www.researchgate.net/publication/11600248_Regulation_of_mRNA_translation> [Accessed 8 May 2020].
13. Van Lint, S., Heirman, C., Thielemans, K. and Breckpot, K., 2013. mRNA. *Human Vaccines &Immunotherapeutics*, [online] 9(2), pp.265-274. Available at: <https://www.ncbi.nlm.nih.gov/pmc/articles/PMC3859745/> [Accessed 8 May 2020].
14. Van Lint, S., Heirman, C., Thielemans, K. and Breckpot, K., 2013. mRNA. *Human Vaccines &Immunotherapeutics*, [online] 9(2), pp.265-274. Available at: <https://www.ncbi.nlm.nih.gov/pmc/articles/PMC3859745/> [Accessed 8 May 2020].
15. Yan, X., Hoek, T., Vale, R. and Tanenbaum, M., 2016. Dynamics of Translation of Single mRNA Molecules In Vivo. *Cell*, [online] 165(4), pp.976-989. Available at: <https://www.cell.com/cell/pdf/S0092-8674(16)30477-9.pdf> [Accessed 8 May 2020].
16. n.d. *Principles Of Biology*. [ebook] Available at: <https://dls.ym.edu.tw/course/hb/doc/lecture6-(18)%2050-Translation.pdf> [Accessed 8 May 2020].
17. Allen GS, Zavialov A, Gursky R, Ehrenberg ,Frank J, 2005
18. Baltimore, 1970
19. Berg, 2002
20. Brenner et al., 1961
21. Browning, 1996
22. Chapman and Brown, 2004
23. Clark and Pazdernik, 2013
24. des Georges et al., 2014

25. Dever and Green, 2012
26. Doerfel et al., 2013
27. Gerlinger et al., 1977
28. Guhaniyogi and Brewer, 2001
29. Gutierrez et al., 2013
30. Hinnebusch et al., 2016
31. Hiraga and Yanofsky, 1972
32. Hwang et al., 2013
33. Jackson et al., 2012
34. Kozak, 1983
35. Lauer et al., 1984
36. Leprivier et al., 2013
37. Li and Nagy, 2011
38. Li et al., 2009
39. Lopez-Valenzuela et al., 2004
40. Matsuda et al., 2004
41. Meijer and Thomas, 2002
42. Nanda et al., 2009
43. Ortiz et al., 2006
44. Petsch et al., 2005
45. Pisarev AV, Hellen CU, Pestova TV, 2007
46. Pisarev et al., 2010
47. Roy and von Arnim, 2013
48. Sasikumar et al., 2012
49. Sasvari et al., 2011
50. Temin and Mizutani, 1970
51. Ude et al., 2013
52. Wang and Krishnaswamy, 2012
53. Zhou et al., 2010b

6

Ribosome Structure and Function

CONTENTS

The chapter focuses on the structure of the ribosome and what is the function of the ribosome. In the first half the chapter, it discusses the parts of ribosomes including its size, composition and general structure of the ribosome. Next, the chapter provides insights about the conservation of structure including its archaebacterial ribosomes and eukaryotic ribosomes. In the second half, the chapter highlights the techniques that can be used to study the ribosome structure which include neutron scattering, electron microscopy, x-ray crystallography and model building. This chapter also provides knowledge about the function of a ribosome.

6.1. INTRODUCTION

There have been often seen that there are many organelles which tend to work together to complete the cell activities while examining the animal as well as plant cell by using a microscope. This is one of the essential cell organelles which are ribosomes. These are in charge of the protein synthesis.

The ribosome is considered to be a complex made of protein and RNA. The ribosome tends to add up to numerous million Daltons in size and it is assumed to have an important part in the course of decoding the genetic message which is reserved in the genome into protein.

There is an essential chemical step of protein synthesis and that is peptidyl transfer. In this, the movement of the developing or nascent peptide takes place from one tRNA molecule to the amino acid along with another tRNA. While developing the polypeptide, the amino acids are included along with the arrangement of codons of an mRNA. Therefore, the ribosome has important sites for one mRNA and not less than two tRNAs.

A ribosome is said to be made up of two subunits. These include the big and the little subunit which contains a couple of ribosomal RNA (rRNA) molecules and a random number of ribosomal proteins. Several protein factors catalyze the distinct impressions of protein synthesis. It is important to consider the translation of the genetic code for the manufacturing of the protein which is useful and also, for the growth of the cells.

Ribosomes are said to be the particles that are present in large numbers in all the living cells. It serves as the site where the protein synthesis takes place. Ribosomes can occur both as free particles in prokaryotic as well as eukaryotic and also, it acts as the particles which are attached to the membranes of the endoplasmic reticulum in eukaryotic cells.

Ribosomes are the small particles which are remarkably abundant in cells. For example, a single actively replicating eukaryotic cell can consist of around 10 million ribosomes. Considering the bacterium *Escherichia coli*, which is a prokaryote, it can be said that the ribosomes can have around 15,000 ribosomes, which is as much as one – quarter of the total mass of the cell. There is variation in the size of the ribosomes within cells, and this depends on the cell type and one some factors that include whether the cell is resting or replicating. There is a best – characterized example and that is of the average ribosome of *E.coli*. This measures about 200 angstroms which is around 20 nm in diameter.

Ribosomes are said to be made up of ribosomal proteins and ribosomal RNA (rRNA). Considering the case of prokaryotes, it can be said that the ribosomes are approximately 40 percent protein and 60 percent rRNA. Considering the case of eukaryotes, it can be said that the ribosomes are approximately half protein and half rRNA. Usually, the ribosomes are made up of three or four rRNA molecules and anywhere from around 40 to 80 different ribosomal proteins.

Each ribosome consists of two subunits which include a larger one and a smaller one. Each of these has a characteristic shape. Typically, the subunits are known in terms of their sedimentation rate. This sedimentation rate is measured in Svedberg units (denoted by S), in a centrifugal field. The small and large subunits of eukaryotes are denoted by 40S and 60S, respectively; whereas, the prokaryotes consist of a small 30S subunit and a large 50S subunit.

Ribosomes can be defined as the sites at which information that is carried in the genetic code is converted into protein molecules. Ribosomal molecules of mRNA that is messenger RNA tend to determine the order of tRNA that is transfer RNA molecules which are bound to nucleotide triplets (referred to as codons).

Finally, the order of tRNA molecules tends to determine the amino acid sequence of a protein. The molecules of rRNA catalyze the peptidyl transferase reaction, which in turn forms peptide bonds in between the amino acids, to form proteins. The newly formed proteins tend to detach themselves from the ribosome site and further, migrate to other parts of the cell to be used.

6.2. PARTS OF THE RIBOSOME

The yeast (*Saccharomyces cerevisiae*) is the canonical model. The structural dynamics of the ribosome have been elucidated upon this.

Figure 6.1. Yeast - *Saccharomyces cerevisiae.*

Source: Image by Wikimedia Commons

The yeast produces around 2000 proteins per minute to satisfy the demand of protein. Each ribosome consists of four rRNAs ribosomal RNAs and 79 r-proteins (ribosomal proteins) which are said to be important to produce a single functioning ribosome particle. Therefore, it is important to have an appreciable amount of energy to make sure that all ribosomes have been produced correctly.

Considering all the species, the characterization of the ribosomal subunits are done by the Svedberg units, that is abbreviated S. These units depend on the rate at which subunits sediment in a solvent; concomitantly, derivation of this is done from the measurement of a parameter known as the apparent sedimentation coefficient.

The simplest ribosome is from bacteria. This ribosome contains the 30S and 50S subunit which altogether forms the 70S ribosome. In contrast to this, the yeast ribosomes contain the 40S and 60S subunit, which come together to form the 80S structure.

There are three sites in each subunit across species which tend to underpin the translation process. These are designated the A that is aminoacyl, which tends to accommodate the incoming tRNA. This tRNA is primed with an amino acid named as aminoacylated-tRNA; P (peptidyl) (that holds the tRNA with the emerging, or nascent, peptide chain that is peptidyl-tRNA); and E (exit), which tends to mobilize the emerging tRNA which has been stripped of its amino acid that is deacylated-tRNA.

As compare to the bacterial ribosomes, eukaryotic ribosomes are large in size as well as more complex. They consist of ES that is ZAQ1 and also, additional r-proteins and r-protein extensions. They consist of over 5500 nucleotides of rRNA which are distributed in the small subunit (18S rRNA) and large subunit (5S, 5.8S, and 25S rRNA) alongside the 79 r-proteins (80 in other eukaryotes).

The crystal structures are followed to complete the full assignment and mapping of rRNAs and r-proteins. These crystal structures are obtained from both the small as well as large subunits *Tetrahymena thermophila* and the *Saccharomyces cerevisiae* 80S ribosome.

The organization of the RNA within ribosomes is done by tertiary structural motifs, which include pseudoknots, which is responsible for carrying out the catalytic activity of the ribosome. The proteins surround the ribosome on the outside and are said to stabilize its structure.

6.3. SIZE, COMPOSITION AND GENERAL STRUCTURE

6.3.1. Size and composition

The ribosome is around globular. It has an average diameter which ranges from 2.5 nm (Escherichia coli) to 2.8 nm (mammalian). The small subunit contains one RNA and between 21 (*E.coli*) and 42 (rat liver) proteins; whereas on the other hand, the large subunit contains two or three RNAs, along with the number of proteins which ranges from 34 (*E.coli*) to 46 (rat liver).

6.3.2. General structure

The construction of all known ribosomes is done of a somewhat flat and elongate small subunit and a somewhat hemispherical large subunit. This associate at the initiation step and dissociate upon the termination of synthesis of the protein.

The ribosome is considered to be a physical scaffold, which helps in bringing together the necessary components including mRNA (which represent a transcript of the DNA segment that is to be translated); tRNA (which bears a specific amino acid and recognition site for the corresponding mRNA codon); GTP that is guanosine triphosphate. GTP is said to be a molecule which supplies energy and a host of protein factors that operate during the process of initiation, elongation and termination.

The most important point is that it has catalytic activities which are located in the peptidyltransferase center. These are required for the transfer of the tRNA-bound amino acid to the nascent polypeptide in every step of elongation. The small subunit (30S) shows a one third or even two-thirds distribution of mass into a head and body portion.

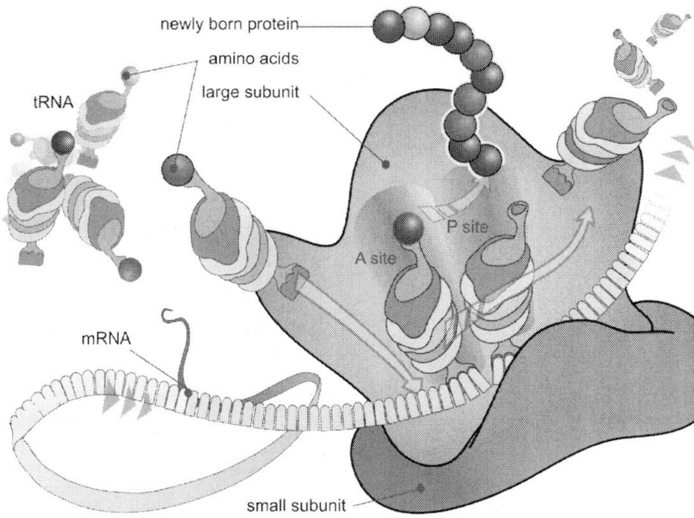

Figure 6.2. Ribosome mRNA translation.

Source: Image by Wikimedia Commons

The upper part of the body tends to bear a broad shelf - like protrusion which is known as the platform. The platform extends along with the interface of the large subunit. The somewhat triangular head tapers off into an extension that is scythe-shaped, curved away from the large subunit, which is specifically pronounced in eukaryotes and is known as the beak, or bill.

The large subunit (50S) is somewhat hemispherical and tends to bear three prominent extensions which being considered to serve as landmarks. Considering *E.coli*, the central extension is known as the central protuberance (CP) is majorly formed by 5S RNA, whereas the identification of the flanking ones has been done by immunoelectron microscopy as protein L1 and a complex which is formed by proteins L7 and L12 ('L7/L12 stalk', St), respectively.

The L7/L12 stalk in *E.coli*, and its analogue in ribosomes from archaea and eukarya, is considered to be highly mobile. It is said to play an active role in the process of binding and movement of tRNAs during the process of elongation. The interface surface of the subunit is relatively flat, despite this, it is interrupted by a deep groove termed IC that is the Interface Canyon.

The interface canyon deepens into a tunnel. This traverses the subunit to exit on its back. The nascent polypeptide chain is threaded through this tunnel to exit from the ribosome. The tunnel appears to provide a tightly sealed compartment which is from the site where the synthesis of the polypeptide takes place, to the point where it exit.

Considering the eukaryal cytoplasmim, it is said that the ribosomes are engaged in the synthesis of exported proteins, the extension of the sealed environment takes place through the protein export channel to which the ribosomes are attached. In their mutual association, the subunits tend to form a space that is partially open and tent-like, and is known as the intersubunit space. This in turn provides the sites of interaction between the ribosome and tRNA molecules.

The sites where the interaction between the codon– anticodon takes place are in the region of the cleft between platform and head of the small subunit; whereas, there are those which are associated with the catalysis of the peptide bond and those are located in the interface canyon of the large subunit, having close proximity to the tunnel entry.

During the process of protein biosynthesis, the tRNA tends to traverse the inter - subunit which is done by going through three primary binding positions, known as the A site (this is the binding site of the aminoacylated tRNA), the P site (this is the site of peptidyl tRNA or the tRNA bound to the polypeptide) and the E site (this is the exit site, where the movement of the deacylated P-site tRNA takes place after the formation of the peptide bond). At any point of time, only two of these sites are occupied which can be A and P (in the pretranslocational state) or P and E (in the posttranslocational state).

6.4. CONSERVATION OF STRUCTURE

The structure of the ribosome is also conserved along with the remarkable functional conservation of the ribosome and also, its principal ligands (including mRNA, tRNA and protein factors) considering the whole kingdom. Even though there are great increases in the translational control and fidelity in higher eukaryotes which include some additional components and steps, there are basic functions of initiation, elongation and termination which closely resemble those in *E.coli*. In addition to this, the topography of the inter-subunit space appears somewhat identical in archaeal, eubacterial and eukaryotic ribosomes.

Therefore, among the ribosomes, the inter – kingdom differences tend to be the most notable in peripheral regions, which are away from the inter – subunit space. There is a head of the small subunit of eukaryotic ribosomes which has a pronounced beak and also, its body has additional lobes (feet) at its bottom.

Considering the higher eukaryotes, there is evolution of the large subunit from a hemispherical to a more hemi-ellipsoidal shape. In general terms, these are the differences that involve the insertions of RNA segments and additions of the proteins in comparison to the ribosomes of eubacteria.

6.4.1. Archaebacterial ribosomes

The ribosomes which are from archaebacteria seem to be intermediate in size and structure in between those of the eubacteria and the eukaryotes. In context to size, they can resemble eubacterial ribosomes much more closely. Despite this, the sequencing studies of their small subunit rRNAs provide further indications about their closer affinity with eukaryotes.

Considering the different branches of archaebacteria, for most of the part, the methanogens and halophiles tend to show closer affinities to the eubacteria in ribosome structure, whereas the methanococcales and sulfolobales tend to more closely resemble with the eukaryotes.

6.4.2. Eukaryotic ribosomes

Among all, there are many eukaryotes that contain two distinct types of ribosomes namely cytoplasmic and organellar. A marked variation can be seen in the cytoplasmic ribosomes along with the complexity of the organism.

The yeast ribosome is said to be relatively small in size and globular in structure and further resembles a eubacterial ribosome excluding the ones with greater size; whereas, on the other hand, there are higher eukaryotes that have larger and more ellipsoidal ribosomes.

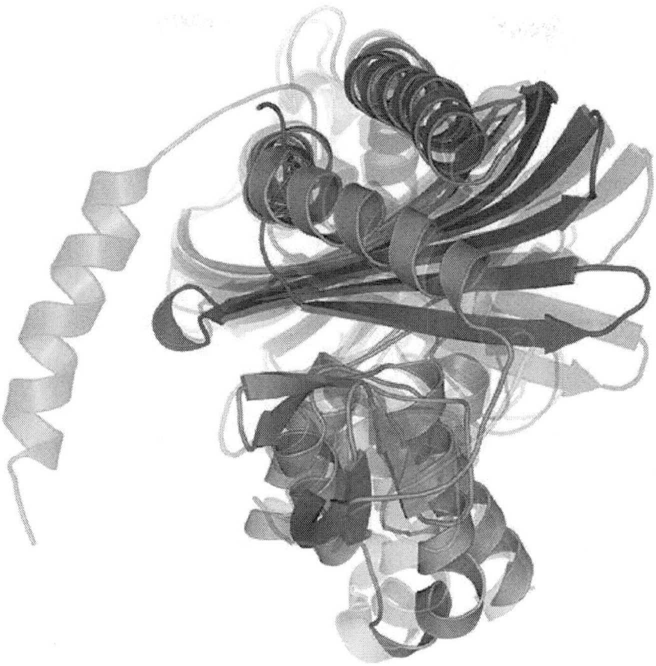

Figure 6.3. Orthogonal views of cytoplasmic capping protein.

Source: Image by Wikimedia Commons

Most of the variation taking place is because of the increases in the size of components of large (60S) subunit. The organellar ribosomes from plastids (which are found in plants only) and mitochondrial ones (which are found in both the animals as well as plants) are said to be small in comparison to the cytoplasmic. Generally, they tend to resemble the eubacterial ribosomes, along with the hypothesis that is these organelles have a eubacterial origin (endosymbiosis).

On one hand, the chloroplast ribosomes have been measured at 67S – 68S, the mitochondrial ribosomes tend to show a wide range of variation that is from 55S (mammalian) to 77S (yeast). However, the structural work on organellar ribosomes lags far behind that for *E.coli* or even eukaryal cytoplasmic ribosomes.

6.5. TECHNIQUES FOR STUDYING RIBOSOME STRUCTURE

There are several techniques which can be used in structural analysis but the main techniques for the analysis of whole ribosomes and their subunits include neutron scattering, electron microscopy and X-ray crystallography.

Along with this, the individual proteins are RNA fragments are continuously studied by using X-ray crystallography and NMR spectroscopy. The three – dimensional models of ribosomal RNA have been obtained by using the model – building further guided by the experimental data and chemical bonding constraints.

6.5.1. Neutron scattering

The main aim of the modern materials science is to understand the factors which tend to determine the properties of matter on the atomic scale and then, to use this information or knowledge for optimizing those properties or for developing the new materials and functionality. This process includes the discovery of physics that is fascinating and also new, which can result in capabilities which were not considered previously.

In society, there have a number of major changes such as the dramatic growth of computing and the internet for the steady increase in the average life span. Almost, all of these changes have their origin while the human is understanding and exploiting the physics and chemistry of materials. Scientists tend to use a number of various tools and techniques which are based on the scattering of beams of particles most of the time, to investigate the atomic – scale structure and dynamics.

There is an 'ideal' probe which can be the one having a wavelength which is similar to the spacing between the atoms for studying the structure with atomic resolution. It can be the one having energy similar to the atoms in materials for studying their dynamics. It would not have charge to avoid strong scattering by charges on the electrons or the nucleus and can allow deep penetration into materials.

The scattering of the same would occur to a similar extent by both light as well as heavy atoms and also, have a suitable magnetic moment so that the study of the magnetism can become easy for the learners too. On an acute scale, the scattering cross – section would be precisely measurable in order to facilitate the comparison with the theory and the computer modeling. This particle which exists is the neutron.

Unfortunately, it is said to be difficult to produce beams of neutrons which are of high-intensity. These high – intensity beams of neutrons are normally only found strongly bound to protons in the nuclei of atoms. Moreover, this can be done by the process of fission in a nuclear reactor. In this, the fundamental process is the release of neutrons that produces heat. A research reactor (for example, that at the Institut Laue-Langevin (ILL) in Grenoble, France) is optimized in order to produce bright beams of neutrons.

There is one more way to produce intense neutron beams and that is by using a source which is accelerator-based. An example can be the ISIS facility near Oxford in the UK, where neutrons are released from tungsten nuclei from a high - energy beam of protons in a process which is referred to as spallation.

Both research reactors and the "spallation sources" are large as well as expensive facilities, therefore, there are relatively few in the whole world. The UK is fortunate enough because they have access to the world's best in each class – ILL and ISIS.

Neutron scattering tends to provide information that is highly complementary to that information which is collected by using other microscopic scattering techniques including those using photons (from visible light to synchrotron X-rays) or electrons (microscopy and diffraction), and also, to standard laboratory measurements. Considering the modern materials science, it is normally the case that there is a need for a variety of techniques in order to tackle any specific problem.

The ILL neutron beams are said to be continuous, whereas, on the other hand, there are those at ISIS which are produced in short bright pulses 50 times a second. This allows for different optimization. Each and every facility tends to operate around 30 separate experimental stations (referred to as instruments), which are tailored individually for a particular type of measurement and the range of scientific applications.

There are several neutron scientists who use instruments at both ILL as well as ISIS. Besides this, there is a large amount of knowledge sharing in between the facilities, along with some advancements in the techniques and instrumentation which benefit both laboratories. This sharing of knowledge tends to extend across the international community, having smaller neutron sources and further, it provides an input to the building of instruments and various national governments which provide funding and expertise to particular projects. The international facilities have complemented the network of European sources at the Japanese Proton Accelerator Research

Complex, and Oak Ridge National Laboratory, US. The neutron scattering in the UK began at the Dido and Pluto reactors at the Harwell Laboratory in the period of 1960s.

The ILL is jointly owned by France, Germany and the UK. ILL has 10 additional scientific member countries and started its operation in the period of early 1970s. It had many modernization programs, further developing new neutron infrastructure and also, introducing concepts related to new instruments.

Since the year 2000, the Millennium Programme of ILL continues to refresh its neutron guide infrastructure and instrument suite, while increasing its effective performance entirely by a factor of 20. ISIS is located at the Rutherford Appleton Laboratory. ILL is owned by the Science and Technology Facilities Council. It started its operations in the year 1984 and it has attracted scientific partnerships from worldwide.

In the year 2009, its scientific capability and capacity for experiments are expanded and a second target station is added along with additional instruments that are further designed to study the soft matter, advanced materials and bioscience.

In this project, the second phase aims to increase the range of capabilities in life sciences, soft matter, neutron imaging and microchip irradiation testing. More than 1000 experiments are completed every year at ILL and ISIS, which cover a wide range of research ranging from pharmaceuticals and healthcare, clean energy and the environment, to materials engineering, nanotechnology, fundamental physics and IT.

These facilities serve an international research community which has more than 4000 scientists. Each produces more than 500 research publications within a year. A future European Spallation Source is going to start operation in the period of 2020s. It is currently being planned by a number of European countries.

6.5.2. Electron microscopy

Electron microscopy (which is abbreviated as EM) is considered to be a technique in order to obtain high resolution images of biological as well as non - biological specimens. Moreover, it is used in biomedical research for the investigation of the detailed structure of cells, tissues, organelles and macromolecular complexes. The EM images are of high resolution which results from the use of electrons. The electrons have very short wavelengths and in EM, it is considered as the source of the illuminating radiation.

Electron microscopy is used along with various ancillary techniques which include thin sectioning, negative staining, immune – labeling. This is done to answer some particular questions. The electron microscopy images provide some important information on the structural basis of cell function and of cell disease.

There are two main types of electron microscope named as the transmission electron microscopy (abbreviated as TEM) and the scanning electron microscopy (abbreviated as SEM). The transmission electron microscope is used for viewing the thin specimens such as tissue sections, molecules, etc. through which electrons can pass while obtaining a projection image. The TEM is said to be analogous in a lot many ways to the conventional (compound) light microscope.

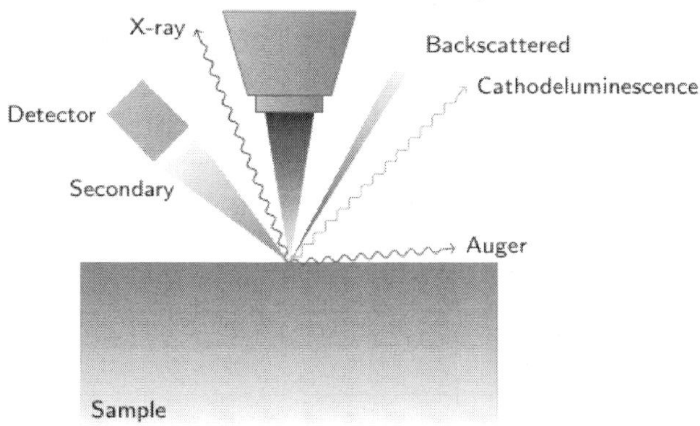

Figure 6.4. Mechanism of SEM.

Source: Image by TeXample.net

TEM is used to image the interior of cells that is within the thin sections, the organization of molecules in viruses and cytoskeletal filaments (that is prepared by using the negative staining technique), the structure of protein molecules (which are contrasted by metal shadowing), and the arrangement of protein molecules in cell membranes (by freeze - fracture).

Negative staining method

An aqueous suspension having purified ribosomes is mixed along with 1 – 2% heavy metal salt for example, uranyl acetate. Then, this is applied to the electron microscope grid. After blotting of the excess liquid is done, the suspension is allowed to dry. The stain then forms a cast around the molecule, which brings out its boundaries having high contrast.

There have been many works which were done earlier such as the study of the morphology of subunits and localization of ribosomal proteins by using the technique of immunoelectron microscopy which were done similarly. Nevertheless, during the grid preparation, the specimen is dehydrated, then the native structure of the ribosome is not well preserved.

Cryoelectron microscopy

There is one more microscopy named as cryoelectron microscopy. In this type of cryoelectron microscopy, the sample is applied to the electron microscopy grid and this is cooled rapidly to liquid-nitrogen temperature. This is done so that the preservation of the macromolecule can be done in a thin layer of vitrified ice under conditions that are somewhat near to those in its native state.

A very low radiation dose is used for visualizing in the transmission electron microscope which is done typically at a magnification of 50,000. When this is done, then the image field shows a number of 'copies' of the molecule that lie in different types of orientations. There can be certain orientations that can occur preferentially which further depends on the nature of the interactions between the molecule and the water surface. On this basis, even they may be close to random.

There are certain mathematical techniques that are used to align and classify the images and also, to find their relative orientations. This kind of usage makes it possible to reconstruct a 3-D image or map of the molecule from these projections. When the molecule images have an extremely low signal – to – noise ratio then it requires collection and processing of several thousand ribosome images.

Till this period of time, among a number of different species (prokaryotes: *E.coli*, *Thermus thermophilus*; archaebacteria: *Haloarcula marismortui*, eukaryotes: *Saccharomyces cerevisiae* (yeast), rabbit reticulocytes and rat liver), the most detailed maps of the ribosome have been obtained in such way, and the mapping of the binding positions of tRNA and elongation factors have been done on the *E. coli* ribosome.

There are many ligands that interact with the ribosome in the course of the translation process, tRNA, initiation factors, elongation factors and release factors. These ligands are large enough so that they can be visualized by using cryoelectron microscopy of the ribosome named as ligand complexes.

When the comparison of such maps is done with control maps then it shows the three-dimensional position of the ligand.

If in case, the X - ray structure of the ligand is known, then in cryo-map, its position can be established with high accuracy by fitting. In order to get success in a ligand localization study, the occupancy of the ligand in the complex must be high (that can be in the range of 80% or above). The reason behind this is the three - dimensional map which is obtained from a number of projections of different ribosomes. But it is difficult to ascertain whether a particular ribosome is occupied or not by using the method of image processing.

Therefore, the mapping of the positions of tRNA on the ribosome has been done in the pre- and post - translocational state. In addition to this, the binding positions of the ternary complex, EF-G, and the Sec 61 channel (for the yeast ribosome) have been obtained.

6.5.3. X-ray crystallography

There is a technique that is used to determining the high-resolution as well as three-dimensional crystal structures of atom and molecules. This technique is referred to as X-ray crystallography. This technique is considered to be fundamental in the development of a number of different scientific fields.

When its first decades of application is considered, then it can be said that it is mainly used to determine the size of atoms, the atomic-scale differences among various materials, the lengths and types of chemical bonds, and also, the crystalline integrity, grain size, grain orientation, film thickness and interface roughness of the materials related, especially minerals and alloys.

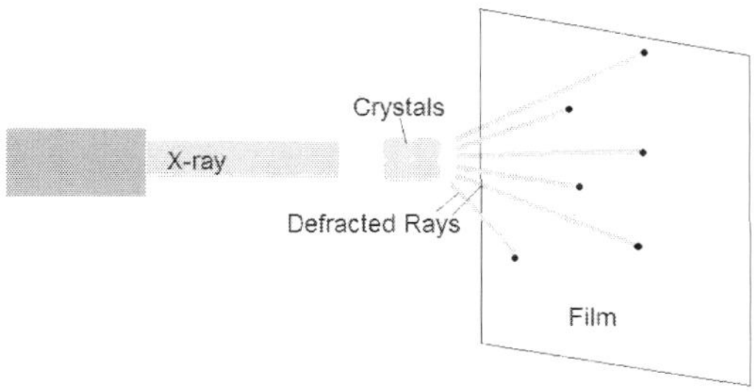

Figure 6.5. X-ray Crystallography.

Source: Image by Wikimedia Commons

This is the method that could also help in revealing the structure and function of several biological molecules which include vitamins, proteins, drugs and nucleic acids. Still, in today's scenario, it is considered to be the chief method to characterize the atomic structure of new materials and in discerning materials that appear somewhat similar by other experiments.

Also, the X-ray crystal structures can explain the unusual electronic or elastic properties of a material. These structures highlighted the chemical interactions and processes, or function and considered then as the basis for designing pharmaceuticals against diseases.

Particularly, by X-ray crystallography, the protein has been extensively put into structure determination. In addition to this, this method is also used in order to determine the way in which a pharmaceutical drug interacts with its protein target and what can be changes which might improve it.

There are some major advantages of X-ray crystallography such as it has no damage to samples, low environmental requirements, free of pollution, high performance and precision over other measuring tools. These are those advantages which make X-ray crystallography the most convenient as well as an important method for investigating the microstructure of materials.

Principle

When a single-crystal X-ray diffraction measurement is considered, then it can be said that in this a crystal is mounted on a goniometer. The goniometer is an apparatus which is used for the positioning of the crystal at selected orientations. The illumination of the crystal takes place with a finely focused monochromatic beam of X-rays. This, in turn, leads to a diffraction pattern of regularly spaced spots which is referred to as reflections.

The technique of X-ray crystallography tends to work in a manner of elastic scattering with the outgoing X-rays which have the energy and wavelength similar to the incoming X-rays. These rays get an altered direction after diffraction.

Then, a three dimensional picture of the density of electrons within the crystal can be produced by a crystallographer. This is done by measuring the angles and intensities of these diffracted beams by further applying the mathematical method of Fourier transforms.

After getting the electron density, the crystallographic disorder, mean positions of the atoms, chemical bonds, and some other information in the crystal can further be determined. There can be some of the poor resolution or even errors which may occur if the crystals are too small, or even if the crystal is not uniform enough in their internal makeup.

Workflow

There are three basic steps in the technique of single crystal X-ray crystallography. The first step is usually the most difficult and that is the production of an adequate crystal of the material to be studied. The crystal must be sufficiently large having all dimensions greater than 0.1 mm, it should be pure in composition and also, regular in structure. They must have no significant internal imperfections including cracks or twinning.

The placing of the crystal is subsequently done in an intense beam of X-rays, and usually, these rays have a single wavelength. This is done to produce regular reflection pattern. The measurement of the angles and intensities of diffracted X-rays are done with each compound while having a unique diffraction pattern. After this, the previous reflections disappear and new ones appear along with the gradual rotation of the crystal. Then, at every orientation of the crystal, the intensity of every spot is recorded.

The collection of a number of data sets have to be done because each set covers slightly more than half a full rotation of the crystal and typically, it contains tens of thousands of reflections. Eventually, these collected data are combined in a computational manner having complementary chemical information to obtain as well as refine a model from the arrangement of atoms present in the crystal. Then, a final refined model of the atomic arrangement is obtained which is referred to as a crystal structure and usually, it is stored in a public database.

6.5.4. Model building

There is a promising approach to ribosomal RNA structure and that is the experimentally guided model building. Considering the ribosome, the RNAs tend to form a distinct structural framework. In this framework, the helical double – stranded segments alternate with single – stranded regions and more complicated structural motifs.

For various species, there are primary sequences of rRNA which are already known. From these, the secondary structure can be inferred as per the rules of base pairing and other atomic interactions, in combination with the results of biochemical studies. Various data from biophysical experiments can be used. These experiments include neutron scattering, chemical cross-linking, fluorescent resonance energy transfer and chemical footprinting.

These experiments help in modeling the three-dimensional folding of rRNA which is present in the small subunit of the *E. coli* ribosome and the modeling is done by several groups. There have some recent approaches to modeling which have also implemented constraints that are provided by the envelopes and internal density distributions of cryo-EM maps.

6.6. FUNCTION OF A RIBOSOME

In any cell, the function of a ribosome is to produce proteins. Proteins are the main components that are used in almost all cellular functions. Some examples can be like when proteins act as catalysts they speed the time of reactions, when they act as fibers then they provide support much more. There are many functions of proteins and some of them are in specific tasks, such as contracting muscle cells. All proteins start as deoxyribonucleic acid (DNA).

There is a special protein named as *RNA polymerase*. This enzyme recognizes sequences in the DNA, tends to bind them by using the other proteins, and it creates a new information molecule. This information can travel from the nucleus to the cytosol of the cell. The strand of RNA(ribonucleic acid) is produced by RNA polymerase. This strand is processed on its way out of the nucleus, and the removal of the areas of the RNA that do not code for proteins is done. Now, the molecule is referred to as a *messenger RNA* or mRNA.

Each mRNA is made up of 4 different nucleic bases which are referred to as nucleic acids. The base pairs are read in series of threes, which further makes up *codons*. Every codon specifies a particular amino acid. The same 20 amino acids are used by all life on Earth. In addition to this, the codons are used to call for those amino acids and are nearly universal.

There is a codon which starts all proteins and is referred to as "AUG". This stands for the sequence of nucleic bases such as Adenine, Uracil, and Guanine. There is a special RNA molecule which can bind to amino acids, and this is termed as a transfer RNA or *tRNA*. The tRNA recognizes this sequence and binds to it. Particularly, this tRNA tends to carry a methionine amino acid. The next amino acid could be any one of the twenty and this depends on the protein that is being built.

This is where the ribosome comes in. While recognizing the structure of the mRNA that is bound to a tRNA, the combination of the two subunits of the ribosome can take place to begin synthesizing the protein from the mRNA strand. The ribosome tends to act as a large catalyst, further forming peptide bonds between amino acids. The tRNA, which is used, is released back into the cytosol. This is done so that it can bind to another amino acid. Ultimately, the mRNA will present a codon to the ribosome which means "stop". The string of amino acids is detached from the last tRNA which is done by the special proteins, and the protein will then be released.

There are different proteins that do require different modifications and transport to a number of several areas of the cell before they function. For example, a ribosome attached to the endoplasmic reticulum will deposit the protein inside which is newly formed. At that place, further modification can take place and then, be folded properly. The formation of other proteins takes place directly in the cytosol where they can begin acting as catalysts for several reactions. All of these proteins that cells need are created by ribosomes. The proteins account for about 20 percent in per cell weigh. There are around 10,000 different proteins in an average cell furthermore, having on average a million copies of each.

It is important to synthesize a lot of proteins. This is the reason for evolution of ribosomes is considered to be an efficient and speedy machine. On average, ribosomes have the ability to add 3-5 amino acids per second to a protein chain. It has been widely known that the largest protein named as *titin*, is around 30,000 amino acids. To synthesize this, ribosomes only take 2-3 hours. There are some short proteins that are made up of only a few hundred amino acids, and these short proteins can be synthesized in minutes.

Once it has been made, the ribosomes cannot shut off. A tRNA is attached with an mRNA with the help of various other proteins, and then, the process of protein synthesis is initiated. There has been many advantages to the viruses because of this. A virus is considered to be a small strand of DNA or RNA. A virus replicates by hijacking the normal machinery of a cell, including ribosomes. The virus uses the ribosomes of a cell to create proteins that are needed for it for the replication of its genome and encapsulating itself so that it can leave the cell. When the genome is injected by a virus into a cell, then, the molecule is treated in the same manner as if the cell had created it. If the virus is DNA-based, then, the DNA makes its way into the nucleus. In the nucleus, the proteins of cell translate it into RNA and this translation is done by ribosomes into proteins. On the other hand, if the virus is RNA-based, then the viral RNA stays in the cytoplasm. Here the interaction with ribosomes can take place directly, further, creating new proteins. In either of the ways, the virus will have the ability to create all the proteins that is required for the replication of its genome and packaging of the copies in new protein capsules. They will become able to travel to a new host cell and then, spread some kind of disease.

6.7. RIBOSOMES AND PROTEIN SYNTHESIS

The process of translation or the process of protein synthesis includes the decoding of an mRNA message into a polypeptide product. Amino acids are strung together covalently and this is done by inter - linking peptide bonds in lengths which range from approximately 50 amino acid residues to more than 1,000.

Every \amino acid has an amino group (NH_2) and a carboxyl (COOH) group. The formation of the polypeptides takes place when an amide (peptide) bond with the carboxyl group of another amino acid is formed by the amino group of one amino acid. Such kind of reaction is catalyzed by ribosomes and it generates one water molecule.

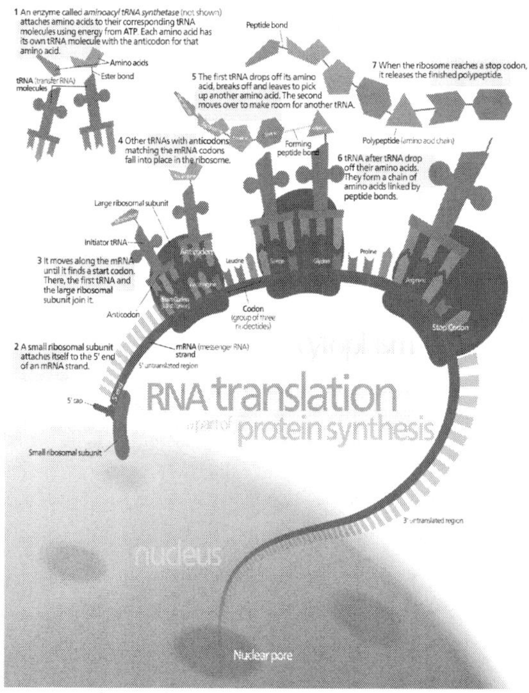

Figure 6.6. Protein Synthesis.

Source: Image by Wikipedia

6.7.1. The Protein Synthesis Machinery

Along with the mRNA template, there are a number of molecules and macromolecules which provide their contribution to the process of translation. There can be variation in the composition of each component across species. For example, the ribosomes can consist of different numbers of rRNAs and polypeptides and this depends on the organism. Nevertheless, the general structures and functions of the machinery of protein synthesis can further be compared with the bacteria to human cells. The translation thus requires the input of an mRNA template, ribosomes, tRNAs, and many different enzymatic factors.

Ribosomes

Prior to the translation of an mRNA, it is important that a cell is investing the energy for building each of its ribosomes. Considering *E. coli*, it can be said that there are ribosomes between 10,000 and 70,000 present in each cell at any specific time. A ribosome can be defined as a complex macromolecule which contains the structural as well as catalytic rRNAs, and a number of different polypeptides. In eukaryotes, the nucleolus is majorly and completely specialized for the synthesis and assembly of rRNAs.

The ribosomes tend to exist in the cytoplasm in the prokaryotes and the cytoplasm and rough endoplasmic reticulum in the eukaryotes. In addition to this, the mitochondria and chloroplasts have their own ribosomes in the matrix and stroma. These ribosomes seem to be very much similar to prokaryotic ribosomes and they have similar drug sensitivities in comparison to the ribosomes just outside their outer membranes in the cytoplasm. The dissociation of the ribosomes takes place into large and small subunits at the time when they are not synthesizing the proteins. Further, the ribosomes re-associate during the initiation of translation. Considering the *E. coli*, the small subunit is described as the 30S, and the large subunit is described as 50S, for a total of 70S (it is important to remember that the Svedberg units are not additives). The ribosomes present in mammals have a small 40S subunit and a large 60S subunit and in a total of 80S. The responsibility of the small subunit is to bind the mRNA template. On the other hand, the large subunit tends to sequentially bind tRNAs.

tRNAs

The tRNAs are said to be the structural RNA molecules which are transcribed from genes by RNA polymerase III. There are 40 – 60 types of tRNAs that exist in the cytoplasm depending on the species. There are specific tRNAs, which serve as adaptors, and they bind to sequences on the mRNA template. Then, they add the corresponding amino acid to the polypeptide chain. This is the reason that the tRNAs are considered to be the molecules which translate the language of RNA into the language of proteins.

Among the 64 possible mRNA codons or the triplet combinations of A, U, G, and C, there are three which specify the termination of protein synthesis and 61 of them tend to specify the addition of amino acids to the polypeptide chain. Among these 61, one codon (that is the AUG) also tends to encode the initiation of translation.

Each and every tRNA anti - codon can form a base pair with one of the mRNA codons. Moreover, according to the genetic code, they can add an amino acid or terminate translation. For example, if on an mRNA template, the sequence CUA occurred in the proper reading frame, then it would bind a tRNA while expressing the complementary sequence that is GAU, which would be connected to the amino acid leucine.

When the tRNAs act as the adaptor molecules of translation, then it is surprising how tRNAs can fit with so much specificity into such a small package. There are three factors which the tRNAs need to interact with. These factors include:

1) they must be recognized by ribosomes
2) they must be recognized by the correct aminoacyl synthetase; and
3) they must bind to the correct sequence in mRNA.

Aminoacyl tRNA Synthetases

There is a process of pre - tRNA synthesis which takes place by RNA polymerase III. This process only creates the RNA portion of the adaptor molecule. Later on, it becomes important to add the corresponding amino acid, at the time when the tRNA is processed and further, exported to the cytoplasm. By the process of tRNA "charging," each and every tRNA molecule is connected to its correct amino acid by a group of enzymes which is termed as aminoacyl tRNA synthetases.

At least one type of aminoacyl tRNA synthetase exists for each 20 amino acids; there can be variation in the exact number of aminoacyl tRNA synthetases which depends on the species. Firstly, these enzymes tend to bind and hydrolyze ATP in order to catalyze a high - energy bond between an amino acid and adenosine monophosphate (AMP). In this reaction, a pyrophosphate molecule is expelled. Then, the activated amino acid is transferred to the tRNA, and the release of AMP takes place.

6.7.2. The Mechanism of Protein Synthesis

Along with the synthesis of mRNA, the protein synthesis can also be divided into three phases which include initiation, elongation, and termination. The process of translation is somewhat similar in the prokaryotes and the eukaryotes.

1. Initiation of Translation

The process of protein synthesis starts with the formation of an initiation complex. Considering *E. coli*, it is said that this complex includes the small 30S ribosome, the mRNA template, three initiation factors (which include IF-1, IF-2, and IF-3), and a special initiator tRNA (which is referred to as tRNAMetf). The interaction between the initiator tRNA and the start codon AUG (or GUG) takes place, and further, it connects to a formylated methionine termed as fMet, and also, it can bind IF-2.

At the beginning of every polypeptide chain that is synthesized by *E. coli*, formylated methionine is inserted by fMet−tRNAMetf; but the fact that it is usually clipped off after the completion of the translation. During translation elongation, when an in-frame AUG is encountered, a non-formylated methionine is inserted by a regular Met-$tRNA^{Met}$.

Considering the *E. coli* mRNA, there is a sequence upstream of the first AUG codon which is referred to as the Shine-Dalgarno sequence (AGGAGG). This upstream interacts with the rRNA molecules which tend to compose the ribosome. This interaction tends to anchor the 30S ribosomal subunit at the correct location on the mRNA template. GTP that is Guanosine triphosphate is a purine nucleotide triphosphate. During the translation both at the start of elongation and during the translocation of the ribosome, this acts as an energy source.

In the case of eukaryotes, there is the formation of a similar initiation complex, which comprises mRNA that is the 40S small ribosomal subunit, IFs, and nucleoside triphosphates (GTP and ATP). fMet in eukaryotes does not get bind by the charged initiator tRNA, called Met-$tRNA_i$, but it is different from other Met-tRNAs in which it can bind IFs.

The eukaryotic initiation complex recognizes the 7-methylguanosine cap at the 5′ end of the mRNA, rather than depositing at the Shine-Dalgarno sequence. A CBP that is cap-binding protein and several other IFs assist the movement of the ribosome to the 5′ cap.

The initiation complex tracks along the mRNA in the 5′ to 3′ direction, further searching for the AUG, once at the cap start codon. The translation of a number of eukaryotic mRNAs is done from the first AUG, but not always. While considering the Kozak's rules, it can be said that the nucleotides around the AUG further indicate whether it is the correct start codon.

According to the statement of the Kozak's rules, '*the following consensus sequence must appear around the AUG of vertebrate genes: 5′-gccRccAUGG-3′*.' Moreover, the R (for purine) indicates a site which can be either A or G, but it cannot be C or U. When the closer the sequence is to this consensus, then the higher will be the efficiency of translation.

Once the identification of appropriate AUG is done, then the other proteins and CBP dissociate, and the 60S subunit tends to bind to the complex of Met-tRNA$_i$, mRNA and the 40S subunit. This is the step that completes the initiation of translation in eukaryotes.

2. Translation, Elongation, and Termination

The basics of elongation are the same in prokaryotes and eukaryotes. There is 50S ribosomal subunit of *E.coli.* This contains three compartments which are named as the A (aminoacyl) site which binds incoming charged aminoacyl tRNAs.

The charged tRNAs are bound by the P (peptidyl) site which carries amino acids. These amino acids have formed peptide bonds along with the growing polypeptide chain but are not yet dissociated from their corresponding tRNA. The E that is exit site tends to release the dissociated tRNAs. This is done so that they can be recharged with free amino acids.

There is one exception which can be considered to this assembly line of tRNAs. In *E.coli*, the fMet−tRNAMetf has the capability of entering the P site directly which is done without first entering the A site. Similarly, the eukaryotic Met-tRNA$_i$ binds directly to the P site with help from other proteins of the initiation complex. Considering both the cases, it can be said that this creates an initiation complex with a free A site which is ready to accept the tRNA that is corresponding to the first codon after the AUG.

During the process of translation elongation, the mRNA template tends to provide specificity. As the movement of ribosome takes place along the mRNA. Each and every mRNA codon comes into register, and specific binding having the tRNA which is corresponding charged anticodon is ensured. If there is an absence of mRNA in the elongation complex, the tRNAs will get bound by the ribosome non - specifically.

The process of elongation proceeds along with the charged tRNAs which enter the A site. After that, they shift to the P site further followed by the E site with each and every single - codon "step" of the ribosome. The steps of the ribosome are induced by conformational changes which further advance the ribosome by three bases in the 3′ direction. For each step, the energy of the ribosome is donated by an elongation factor which tends to hydrolyze GTP.

There are peptide bonds that form in between the amino group of the amino acid which is attached to the A-site tRNA. In addition to this, the bonds are also formed between the carboxyl groups of the amino acid which are attached to the P-site tRNA. The formation of each peptide bond is catalyzed by peptidyl transferase. This is an RNA-based enzyme that is integrated into the 50S ribosomal subunit.

When the ribosome steps across the mRNA, the former P-site tRNA tends to enter in the E site, furthermore, it detaches from the amino acid, and then is expelled. The time taken by the *E. coli* translation apparatus to add each amino acid is only 0.05 seconds. This means that the translation of the 200-amino acid protein can only be in just 10 seconds.

The process of termination of translation occurs when a nonsense codon (which includes UAA, UAG, or UGA) is encountered. When the alignment is done along with the A site, the recognition of the nonsense codons is done by release factors in prokaryotes and also, in eukaryotes. The prokaryotes and eukaryotes instruct peptidyl transferase to add a water molecule to the carboxyl end of the P-site amino acid.

This is the reaction that forces the P-site amino acid so that they can detach from its tRNA, and then the release of the newly made protein takes place. The small as well as the large ribosomal subunits dissociate from the mRNA and also, from each other. Immediately, they areused into another translation initiation complex. After the completion of the translation of a number of ribosomes, the mRNA is degraded so that the nucleotides can be reused in some other transcription reaction.

3. Protein Folding, Modification, and Targeting

During and after the process of translation, there can be a chemical modification in the individual amino acids, furthermore, the signal sequences may be appended, and the new protein gets fold into a distinct three-dimensional structure as a result of intra-molecular interactions.

A signal sequence is defined as a short tail of amino acids that provides directions to a protein to a particular cellular compartment. At the amino end or the carboxyl end of the protein, these sequences can be considered to be the train ticket of protein to its ultimate destination.

There are some other cellular factors that recognize each signal sequence and provide help in transporting the protein from the cytoplasm to its correct compartment.

There are a number of different proteins that fold spontaneously, but there are also some proteins, which do require helper molecules, and these helper molecules are referred to as chaperones. These molecules help in preventing them from aggregating when it is undergoing the complicated process of folding.

Even if there is a case when a protein is properly specified by its corresponding mRNA, then it could take on a completely dysfunctional shape if there is any kind of abnormality in the temperature or pH conditions that can prevent it from folding correctly.

6.8. CONCLUSION

The ribosomes are the small particles which were first described by George E. Palade who was a Romanian-born American cell biologist in the year 1955. He found them to be frequently related to the endoplasmic reticulum in eukaryotic cells.

The structure of ribosome is highly conserved across all species. Despite this, its exact composition tends to depend on the organism. The conclusion that can be drawn from the above chapter is that the major and basic function of a ribosome is to produce proteins. Proteins are considered to be the main components that are used in almost all cellular functions.

The chapter discusses that when the synthesis of proteins takes place then it consumes more energy of a cell in comparison to any other metabolic process. And thus, the proteins account for more mass when compared to any other component of living organisms (the only exception here is water). The proteins tend to perform virtually every function of a cell.

REFERENCES

1. Biology Dictionary. n.d. *Ribosome*. [online] Available at: <https://biologydictionary.net/ribosome/> [Accessed 29 April 2020].
2. Courses.lumenlearning.com. n.d. *Ribosomes And Protein Synthesis | Biology I*. [online] Available at: <https://courses.lumenlearning.com/biology1/chapter/ribosomes-and-protein-synthesis/> [Accessed 29 April 2020].
3. Encyclopedia Britannica. 1998. *Ribosome | Cytology*. [online] Available at: <https://www.britannica.com/science/ribosome> [Accessed 29 April 2020].
4. Frank, J., Agrawal, R. and Verschoor, A., 2001. Ribosome Structure and Shape. *Encyclopedia of Life Sciences*, [online] p.4. Available at: <https://www.researchgate.net/publication/228050512_Ribosome_Structure_and_Shape> [Accessed 29 April 2020].
5. Microscopemaster.com. n.d. *Ribosomes_Definition, Structure, Size, Location And Function*. [online] Available at: <https://www.microscopemaster.com/ribosomes.html> [Accessed 29 April 2020].
6. n.d. *Neutron Scattering*. [ebook] p.2. Available at: <https://www.iop.org/publications/iop/2011/file_47455.pdf> [Accessed 29 April 2020].
7. Shaffer, D., n.d. *Ribosome Structure*. [online] News-Medical.net. Available at: <https://www.news-medical.net/life-sciences/Ribosome-Structure.aspx> [Accessed 29 April 2020].
8. Stock, D., Perisic, O. and Lowe, J., 2005. *X-Ray Crystallography - Creative Biomart*. [online] Creativebiomart.net. Available at: <https://www.creativebiomart.net/resource/principle-protocol-x-ray-crystallography-393.htm> [Accessed 29 April 2020].
9. University of Massachusetts Medical School. n.d. *What Is Electron Microscopy? - UMASS Medical School*. [online] Available at: <https://www.umassmed.edu/cemf/whatisem/> [Accessed 29 April 2020].

7

Inhibitors of Protein Synthesis, Antibiotics and Resistance

CONTENTS

Protein synthesis is regarded as a complicated multifarious process involving numerous enzymes along with conformational alignment. Nevertheless, most of the antibiotics that would obstruct bacterial protein synthesis, would be interfering with the developments at the subunit 30S or subunit 50S corresponding to the bacterial ribosome 70S.

The aminoacyl tRNA synthetases that are supposed to be activating each amino acid needed for peptide synthesis are not considered as antibiotic targets. Contrary to that, the main steps in the process that would be attacked are;

- Forming the 30S initiation complex (including mRNA, the 30S ribosomal subunit and formyl-methionyl transfer RNA)
- Forming the 70S ribosome with the help of 30S initiation complex and the ribosome 50S
- Amino acid assembling within a polypeptide through elongation process

The ribosome would be considered as one of the chief targets in the cell for antibiotics use clinically. But, with the augment of multidrug resistant bacteria, the efficiency of the present ribosome targeting antibiotic arsenal is supposedly being fast reducing. Hence, it would highlight the requirement for discovering elements that would contain new scaffolds for binding the ribosomes with the novel sites.

In this chapter, we will discuss about the protein synthesis inhibitors and tried to emphasis on the various aspects of bacterial protein synthesis inhibitors. Additionally, the chapter highlights the effectiveness of antibiotics in treating serious infections and deadly diseases. The various drugs and their action on the ribosomal cells in reducing infections and diseases would help to know the real function of the antibiotics clinically.

Further, it states the implications of over usage of antibiotics as well as the evolution of drug resistant pathogens. The chapter explains how pathogenic bacteria easily develops antibiotic resistance and points out the risks of using the drugs excessive along with the requirement of new antibiotic drugs which would be able to combat the issue and match with the novel mechanism.

7.1. INTRODUCTION

Studies related to protein synthesis inhibitors' methods of action could be sequentially categorized into two periods. The first period would include the years 1951-1960 and the second , 1961-1973. The purpose of such sequential distinction is due to the fact that in 1951 the precise inhibition of bacterial protein synthesis through antibiotics (chloramphenicol and chlortetracycline) was first described (Gale et al., 1951, Hahn et al., 1951).

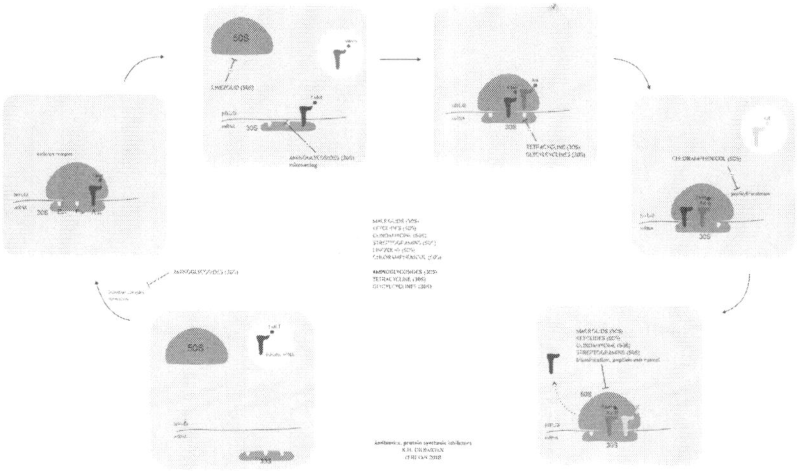

Figure 7.1. Inhibitors of Protein Synthesis

Source: Image by Wikipedia

Moreover, in 1961, it was established through a ribosomal amino acid inclusion system guided by a particular polynucleotide as mRNA (Nirenberg, M.W. and Matthaei, J.H. 1961). From 1961 onwards, studies concerning antibiotic actions mechanisms started developing extensively (Brock, 1961, Gale, 1963, Newton (1965). Thus, covering the subject in a single contribution had become almost implausible, leading to analyzing of protein synthesis inhibitors through independent contribution as well as in complete volumes (Newton et al., 1966, Gottlieb et al., 1967, Weisblum et al., 1968, Pestka, 1971, Muiioz et al.,1972, Cundliffe, 1972), for last eight years.

Though translation upholds universal importance, yet, unique differences between eukaryotic and prokaryotic translation exist, which have been historically exploited to develop antibiotics. While treating a bacterial pathogen with a translation inhibitor, protein synthesis would quickly stop. It would be followed by death or extreme growth regulation. The translation would be inhibited through a number of different strategies, as per recent treatments. It would directly start from the ribosome targeting aminoacyl tRNA (-tRNA) synthetases (Tenson T et al., 2006, Wilson DN. 2014, Lin J et al., 2018, Ho JM et al., 2018, Francklyn CS et al., 2019).

It has been an effective method to treat infections and as a result, dozens of various translation inhibitors were introduced to the clinics since the 1940s, successively transforming healthcare (Ventola CL., 2015, Durand GA et al., 2018). Development of antibiotics helped to cure infections, once considered to be deadly, in merely a few days.

However, as soon as the drugs would have been invented, it would even be declared obsolete in no time. It is because of the acquisition and evolution of resistance mechanisms among bacteria from being eradicated. With inordinate and improper antibiotics usage, growing and spreading of multidrug-resistant pathogens, commonly known as superbugs, have increased considerably (Ventola CL., 2015, Charani E et al., 2019).

Designing strategies to curb the evolution of antibiotic resistance would be a significant global challenge concerning the life sciences community and public health. With recent support of discovering new antibiotics categories, the currently treatable infections might become life-threatening all over again. In the last decade a considerable multiplication of one or numerous antibiotics resistant human-pathogenic bacteria, had been witnessed across the globe.

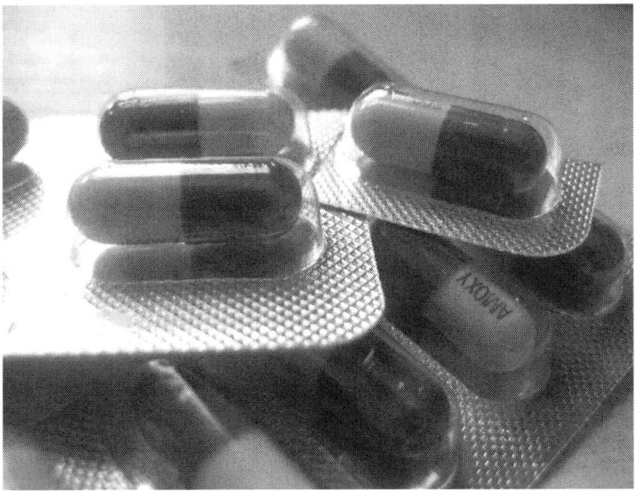

Figure 7.2. Antibiotics

Source: Image by Flickr

Numerous multi-drug-resistant pathogenic bacteria have been developing due to the generous usage of antibiotics in medicine and agriculture. It has resulted in the restraining the effectiveness of the prevalent antibiotic arsenal. The subsequent reduction in precisely new categories of antibiotics getting included in the clinical pipeline would be augmenting the health concerns.

Additional, infections caused by the immune micro-organisms would not respond to mainstream treatment. Moreover, in few of the cases, the antibiotics used as a stopgap, too, would have lost their potentialities. Additionally, over the past decades, the industry records for the evolution of novel antibiotics are considered to have been exhausted.Many of the ribosome-targeting antibiotics which are under clinical trials are supposed to be semi-synthetic by-products of the antibiotics formed out of natural products, are used for clinical purposes till date. Hence, for the antibiotic binding site on the ribosome, they would use an equal scaffold and a comparable interaction mode.A recent world health day had been organized by the World Health Organization with the title "Combat drug resistance: no action means no cure tomorrow". It resulted in stimulating growth in research activities and development of multiple favorable strategies for re-establishing treatment options to fight against infections caused by the resistant bacteria pathogens.The emerging threat of a post-antibiotic world prevailed; however, efforts to evolve new approaches for targeting translation would as well provide an optimistic perspective. An in-depth aspect of the antibiotics action system supposed to target translation and distinct mutations that would be conferring resistance, has been delivered through various current reviews (Tenson T et al., 2006, Wilson DN., 2014, Lin J et al., 2018, Francklyn CS et al., 2019, Ero R et al., 2019, Nguyen F et al., 2014).

The prokaryotic inhibition of protein synthesis through ligands of small molecules has equipped modern medicine with a vital anti-infective compounds arsenal. Also, it contributed in helping to analyze the protein synthesis trail. The collective knowledge acquired through the studies of translation inhibitors would be related to establishing the potential of adopting a chemical biology strategy for the study of the particular process.

The classification of actions which would lead to the assembling of active polysomes have been allowed by the pharmacological inhibitors of protein synthesis. That would even include capturing of the intermediates pertaining to the initiation and elongation cycle as well as a furnishing with the perspectives related to molecular functions of protein factors (Pestka S., 1977, Vazquez D., 1979).

Surprisingly, with most of the compounds being presently used, which are supposed to have been known since many decades ago, reports concerning the documentation of the systematic search of new eukaryotic protein synthesis inhibitors, have been unavailable (Pestka S., 1977, Vazquez D., 1979). The motivation for determining new chemical screens is emerging from the utility of protein synthesis inhibitors as tools, as well as, from the number of studies rising to connect with derestricted translation to contribute to cancer initiation and development (Watkins S.J et al., 2002, Graff J.R. et al., 2003).

For example, malignant transformation can be a result of thefew of the initiation factors being over-expressed. While, the same factors being down-regulated, would be able to overpower the changed phenotype (reviewed by Hershey J.W. et al., 2000). In cancer, the translation device components are considered to be over-expressed or mutated (Hershey J.W. et al., 2000). The translation process is impacted directly followed by the levels of ribosomes being affected, by the tumor suppressor gene product (pRB) (Ruggero D. et al., 2003).

For various cancers, the PI3K/Art signaling route would be uncontrolled and would lead to the modulation of events of factors limiting protein synthesis initiation (Meric F. et al., 2002). Hence translation would represent a fascinating less explored target for chemotherapeutic intervention.

A major part of the identified protein synthesis inhibitors would bind to ribosomes and few among them would even make significant contacts with the RNA component (Brodersen D.E. et al., 2000, Carter A.P. et al. 2000, Pioletti M. et al., 2001, Hansen J.L. et al., 2002). The concept of structured RNA is confirmed by such findings and regards it in the form of biological targets to interdict small molecule. Further, it has been established through recent studies that prokaryotic mRNAs would be able to act as metabolite-responsive genetic switches through their highly structured domains (Mandal M. et al., 2003).

As per the cases studied so far, the metabolite would bind to an RNA motif leading to an allosteric transformation resulting in terminating the transcription process (Mironov A.S. et al., 2002) or initiating the translation process (Nahvi A. et al., 2002). In eukaryotes, the RNA interactions in small molecule (Werstuck G. et al., 1998, Harvey I. et al., 2000) or small molecules in the form of chemical inducers of dimerization (Harvey I. et al., 2002) would be able to impede protein synthesis through translation initiation interference.

The efficiency to regulate gene expression through chemical ligands balancing mRNA function would be left unexplored to a great extent. RNA is considered to be a quality target due to its various features; for several cellular functions it would be considered as vital, would be comprised of intricate secondary and tertiary structural folds and would even exhibit cellular repair deficiencies.

The power of small molecules for interaction with RNA would be well known (Gold L. et al., 1995) and would have the ability to avoid or boost gene expression, acquire allele-specific variation related to gene expression (while allelic sequence variations result in changed RNA conformations), would have isoform-specific gene expression modulation and show impacts of allosteric.

7.2. BACTERIAL PROTEIN SYNTHESIS INHIBITORS

Pathogenic bacteria would easily develop antibiotics resistance. It would generate a continuous requirement for new antimicrobial agents to prevent infections caused through resistant pathogens (Levy SB et al., 2004). Though most of the clinical antibiotics are considered to be either natural products or belonging to their semisynthetic derivatives (Newman DJ et al., 2012), yet, finding the new scaffolds are supposed to have diminished with time (Li J et al., 2009). It would be relatively due to the constant reisolation process of the identified compounds.

It resulted in rejecting several antibiotic discovery programs. Meanwhile, the drug-resistant bacteria would be considered to have been spreading and progressing ahead of lead compounds being developed, competent to treat related infections.

Hence, new technologies are in the process of being developed to enable researchers for accessing natural and synthetic products which were prior ignored by conventional screening programs. In short, one of the strategies for rapid screening of small molecules serving through novel mechanisms of action (MOA) would be the emergence of new antibiotic assays specifically (Peach KC et al., 2013, Schulze CJ et al., 2013, Wong WR et al., 2012).

A substantial number of drugs used for the clinical purpose would obstruct the growth of pathogenic bacteria with the help of a ribosomal bond and protein synthesis interference (Wilson DN., 2009). The ribosome is considered to be a 2.5-MDa multimolecular ribonucleoprotein complex made from small and large subunits. Several opportunities for inhibition of its activity would be provided by the functional complexity of the ribosome.

Several components of protein synthesis encounter interference from the antibiotic binding at particular functional centers. It would consist of, however, not limited to tRNA binding, mRNA translocation, the formation of the peptide bond and exit of the nascent peptide. Hence, identifying the exact site and means of actions of a new inhibitor would be tough due to such complexities.

Generally, the analysis would be inclusive of a lengthy series of sophisticated assays that would require separate purified components of the translation machinery. It would further include testing the inhibition of particular ribosomal activities or meticulous characterization of the site of the drug action.

Thus, the presence of additional general biochemical assays proficient in quick identification of the MOA related to novel inhibitors, are supposed to be capable of considerably simplifying the discovery process initiated by the drug.

Essential evidence provided by genetic approaches about protein synthesis inhibitors would frequently act as a supplement in biochemical MOA investigation. Direct rRNA interaction with the majority of ribosome-targeting inhibitors (reviewed by Poehlsgaard J et al., 2005) results in the isolation of resistant mutants with nucleotide transformations in rRNA. It is considered to be the most conclusive technique of understanding the site of antibiotics binding to the ribosome, at the same time, the translation interference mannerism. Sander

Nevertheless, due to the plethora of rRNA genes in bacterial genomes, isolating a mutant developing resistance through rRNA mutation would be difficult. Going by the clinical perspective, it would be highly valuable as resistance would be seldom related to rRNA mutation; however, it would greatly obstruct experimental identification of the drug action sites.

Selecting resistant mutants through rRNA changes have been conducted traditionally with the help of specialized model organisms which would either have one rRNA allele naturally (e.g., *Halobacterium halobium* [Hummel H et al., 1987]) or devised for conducting a single replica of rRNA genes the respective chromosomes (e.g., *Thermus thermophilus* [Hummel H et al., 1987]) or *Mycobacterium smegmatis* [Sander P et al., 1996]).

Despite such experimental organisms being valuable for categorizing the sites of action for several antibiotics, a maximum number of such models would have downsides. The archaeon *H. halobium* is supposed to be growing sluggishly and would show impassiveness to various antibacterial as a result of the consequential evolutionary separation of bacterial and archaeal ribosomes. The strain engineered from *T. thermophilus* accompanying a sole rrn allele (Gregory ST et al., 2009, Monshupanee T et al., 2008) would require an adversely high growth temperature (ca. 65°C) which would lead to compromising the stability of several antibiotics.

The operon form in single rRNA belonging to *Mycobacterium smegmatis* (Sander P et al., 1996), despite being instrumental in clarifying the means of action pertaining to vital antibiotics (Sander P et al., 2002, Springer B et al., 2001), would relatively grow at a sluggish pace and the outer cell wall layer would have unusual lipids that result in damaging the process of absorption of few drugs (Lambert PA., 2002).

Easy isolation pertinent to a consistent community of the resistant ribosomes towards successive biochemical characterization would not be allowed by means of additional experimental methods using plasmid-borne rRNA genes (Cochella L et al., 2004, Ettayebi M et al., 1985). Hence, through the access of a potent and biochemically docile model for isolation of rRNA resistance mutations is considered to extremely accelerate the investigation of new antibiotic leads.

Therefore, for quick identification of the binding site and MOA related to ribosome-targeting antibiotics, a two-pronged method known as ChIPS (Characterization of Inhibitors of Protein Synthesis) have been described. By combining the two assays including ChIPS would allow for the quick characterization of unidentified protein synthesis inhibitors. Further, it would facilitate in spotlighting the known successes through high performance screening in association with natural product extracts or chemical libraries.

First, two antibiotic-hypersensitive *Escherichia coli* strains would be engineered that would have a single rrn allele and could be applied for quick isolation of resistance mutations as well as pinpointing the ribosomal site of antibiotic action. It would be followed by the demonstration of a cell-free translation pattern where antibiotic-triggered ribosome would interrupt mRNA showing inhibitor-specific, idiosyncratic structures.

7.3. INHIBITORS OF PROTEIN SYNTHESIS

The topic would include varied groups of antibiotics categorized together due to their common action mechanism; i.e., all would be included in the category of protein synthesis inhibitors. The targeted drug for all such different levels of drugs would frequently differ (though some would overlap); however, each results in impacting protein assembly at the ribosomal position.

Following would be regarded as the different types of antibiotics: macrolides/lincosamide, aminoglycosides (the sole bactericidal type in the group), tetracyclines, chloramphenicol, streptogramins and oxazolidinones. Such drugs are seldom regarded as drugs of choice either for a specific infection or for a specific bacterial species. However, they would consistently acquire a critical role as antibacterial agents. A general analysis of protein synthesis inhibitors (translation) is as follows:

7.3.1. Macrolides/Lincosamide

The macrolides belong from a family of harmless, bacteriostatic drugs which would be commonly used for treating infections acquired through the community. Erythromycin considered as the original representative and would be used prevalently in various iterations. Azalides, being the second generation of such drugs, would possess a broad spectrum of activity along with diverse pharmacokinetics. Lincosamide restricted to clindamycin, is a drug that would majorly be used for its anaerobic activities.

Chemical Structure:

The large structure of lactone ring comprising of 14-16 members.

Mechanism of Action:

This would bind with the 50S ribosomal subunit in a reversible manner obstructing the tRNA binding to the acceptor site. It would result in obstructing the translocation of the peptide chain. Since *in vitro* studies would exhibit interference of macrolides with the binding of other protein synthesis inhibitors, such drugs (e.g., chloramphenicol), are supposed to contain overlapping binding sites.

Mechanism of Resistance:

In recent times, usage of such type of drugs is considered to have been limited due to resistance. Bacteria might have macrolide resistance capabilities due to a permeability obstruction (Enterobacteriaceae). The drug could be expelled in Gram-positive species through active efflux pumps. Dominant macrolide resistance in few species could be conferred through a one-step mutation in 50S rRNA genes. Ultimately, the binding of erythromycin would decrease by the 50S rRNA subunit being modified. It can be attributed as a plasmid mediated function.

Antibacterial spectrum:

Macrolides are considered to be active in contrary to Gram-positive bacteria and that would include streptococci, pneumococci, and methicillin sensitive staphylococci. The restricted Gram-negative spectrum would be inclusive of Bordetella and campylobacter. The macrolides would even show activity opposing to mycoplasma, legionella, chlamydia and the treponemes.

Azalides would contain a higher level spectrum including activity opposing to Hemophilus species, Mycobacterium avium intracellular (MAI) and would be dependent on higher azalide activity against Gram positives (clarithromycin) or Gram-negatives (azithromycin).

Pharmacology:

Macrolides are usually meant for oral intake. Since they are not regarded as acid stable, usually they would be made with a coating of acid resistance. In case of further serious infections, intravenous provisions would be available. The t1/2 is at 1.4h and considering azalides, it would be greatly extended making room for once or twice therapy daily. Such group of agents would gain effective distribution, and notably, would gain excessive intracellular concentrations (specifically the azalides).

Toxicity:

Toxicity is considered to be restricted. In macrolides, the GI symptoms would be more regular than azalides.

Indications for use:

Community-developed pneumonia-mycoplasma, legionella, chlamydia infections; pertussis, campylobacter gastroenteritis and MAI. The usage of such drugs is meant for an investigative purpose in the form of anti-inflammatory agents in cystic fibrosis and for preventing atherosclerotic cardiovascular disease.

7.3.2. Aminoglycosides

The aminoglycosides would be regarded as bactericidal agents used for treating severe infections. Due to apprehensions related to the development of resistance. Additionally, they possess a very limited therapeutic to a toxic ratio (i.e., the serum level is required to be powerful being a therapeutic agent would be near to the level where the toxicity of the drug would be met).

Chemical Structure:

Comprises of minimum two amino sugars connected through glycosidic bonds to an aminocyclitol ring.

Mechanism of Action:

A fast bactericidal agent in association with numerous mechanisms of action. The 30S ribosome would be considered as the primary target that would initiate the termination of premature chain and misreading of the RNA codon. Additionally, it would result in the leakage of the outer membrane of Gram-negatives. The grounds for bactericidal activity have not been comprehended entirely.

Mechanism of Resistance:

Enzymes being modified by aminoglycosides would be the primary channel of resistance. Adenyl transferases or phosphotransferases enzymes would help in modifying the aminoglycoside making them ineffective. In such enzymes, the genes are usually plasmid borne and it would result in them getting easily transferred among various bacterial species. Reduced absorption of the aminoglycoside through mutations in the electrochemical gradient would be an added resistance mechanism.

Antibacterial spectrum:

Aminoglycosides would have a rather restricted spectrum of activity. They would seem to be active against Enterobacteriaceae, pseudomonas species and Gram-positive bacteria with staphylococci and streptococci. Few among the aminoglycosides would be active in opposition to mycobacteria. Their activity against anaerobes would be nil.

Pharmacology:

Aminoglycosides would majorly serve as parental therapy. Due to its low protein binding and water solubility qualities, there would be fair distribution into tissues. The infiltration through the blood-brain barrier would be equally poor as it would be in case of bronchial secretions infiltration.

Aminoglycosides would be eliminated in the form of waste through the urine, almost unchanged.

Toxicity:

The level of toxicity in such drugs would be excessive. That would be inclusive of nephrotoxicity in 5-25% of patients, ototoxicity coupled with cochlear and vestibular damage and rarely, neuromuscular obstruction. The drug levels would be usually observed in case of renal disease subjects, to lower the risks such complications.

Indications for use:

Combination of empirical therapy intended for life-threatening infections. Used in the form of combination therapy meant for resistant bacterial infections and combination (synergy) therapy used for enterococcal infections that would necessitate bactericidal activity.

7.3.3. Tetracyclines

The tetracycline type of antibiotics is supposed to be belonging from a huge family of broad spectrum, bacteriostatic agents consisting of limited, however, vital role in the armamentarium of bacterial agents. In 1950s, tetracycline is considered to be the product of a soil-screening program to detect antimicrobials.

Chemical Structure:

The primary structure would be consisting of four bonded 6-carbon rings. The activity of various members of the type would be created through modifying at designated carbon sites.

Mechanism of Action:

The 30S ribosome would be the major target which would prevent the binding of aminoacyl tRNA to the acceptor (A) site on mRNA.

Mechanism of Resistance:

The chief mechanism of resistance to tetracyclines would be low or high export of the drug through efflux pumps. The mechanism might be considered as plasmid mediated as well as gives resistance to every member of the group.

Antibacterial spectrum:

Though such drugs would have Gram-positive activity, inclusive of pneumococcus, streptococci and enterococci, yet, they would be seldom used due to their resistance. They would possess activity in opposition to Gram-negatives – *E. coli*, however, excluding other Enterobacteriaceae, Neisseria, Hemophilus as well as few Shigella. Even they would be active against mycoplasma, rickettsia, legionella, spirochetes and chlamydia.

Pharmacology:

Maximum tetracyclines would be given orally. Both short and long acting agents would exist. A greater number would include lipophilic with brilliant tissue distribution. The most frequently used tetracyclines, like minocycline (metabolized in the liver) and doxycycline (inactivated in the intestine), would be removed through non-renal tracks.

Toxicity:

Though, the toleration level is usually good, yet, it would have an index of major side effects. That would be consisting of photosensitivity, discoloration of teeth in children, hepatotoxicity and hypersensitivity reactions.

Indications for use:

The major role would be to treat Lyme disease, like community-developed pneumonia and acne.

7.3.4. Chloramphenicol

The drug majorly being of historical concern, but since its use is prevalent with a certain frequency in the evolving world, it would be advisable to be recommended. It would be regarded as a wide spectrum agent specifically a sole member of its group. It would gain substantial concentrations resulting in oral administration being included in CSF.

It would act opposing meningitis -causing agents, like, pneumococcus, Hemophilus and Neisseria. It would bind to the 50S ribosome that would avoid linking to the acceptor (A) site on mRNA. Chloramphenicol is supposed to be causing a dose related marrow depression and barely aplastic anemia that would be idiosyncratic. The Grey baby syndrome was supposed to be a matter of concern in neonates.

7.3.5. Streptogramins

It would be a two drug combination originated from pristinamycin. The first would be quinupristin and the second would be dalfopristin. Functioning at the 50S ribosomal subunit, both would be regarded as protein synthesis inhibitors. The combination would be solely used for the treatment of resistant Gram-positive infections which would include vancomycin resistant enterococci (only *E.faecium*) as well as complex methicillin- resistant *S. aureus* (MRSA) infections where vancomycin would be ineffective.

Ribosomal modification (methylation) mainly leads to developing resistance. It could be by means of plasmid mediation. The drug would be considered bactericidal if the isolate is prone to a macrolide. It would be availed only parenterally. Phlebitis at the site of infusion and myalgias which could be rather troublesome would be included as the toxicity levels.

7.3.6. Oxazolidinones

Linezolid is supposed to be the single member of the new group of bacteriostatic agents. The aforementioned family of drugs would be active at the subunit of 50S ribosome. It would inhibit the development of the initiation complex. It would be active in opposition to Gram-positive bacteria, inclusive of MRSA and every vancomycin -resistant enterococci.

Going by its pharmacological benefits, the drug would be available in both oral (with high absorption) as well as parenteral form. Mutation related to the ribosomal binding site would result in the occurrence of resistance. The drug is considered to be properly circulated and effectively tolerated. Prolonged administration (more than two weeks) of the drug would lead to thrombocytopenia and myelosuppression.

7.4. ANTIBIOTICS TARGET THE SMALL SUBUNIT OF THE RIBOSOME

7.4.1. Antibiotics Binding the Decoding Centre

The region encompassing the A site on the 30S subunit which would be observed the actual base pairing among the mRNA codon and the anticodon associated with aa-tRNA, is considered to be the decoding centre (DC). The ribosome associated with this part would contain regions of the 30S head and shoulder, and the top of helix 44 (h44) of 16S rRNA.

It has been shown through biochemical (Abdi NM et al., 2005, Yoshizawa S et al., 1999, Powers T, et al., 1990) and structural studies (Carter AP et al., 2000, Ogle JM et al., 2001) that conserved nucleotides, like A1492 (h44), A1493 (h44) and G530 (h18) would play crucial roles in the process of decoding. Cognate binding of the tRNAs in the A site would let nucleotides A1492 and A1493 to withdraw from h44. It would result in the formation of A-minor interactions monitoring the geometry of the first two codon-anticodon base pairs and confirm the acceptance of Watson-Crick pairs only (Ogle JM et al., 2001).

The ribosomal protein uS12 (Pro45) and nucleotide G530, would monitor the base pair geometry appearing in the third position, in a bit relaxed manner. That would let the vibrating base pairs in that particular position (Ogle JM et al., 2001). Antibiotics related to many different groups of chemical would bind in the DC, and few among them would be used for treating a wide range of infections.

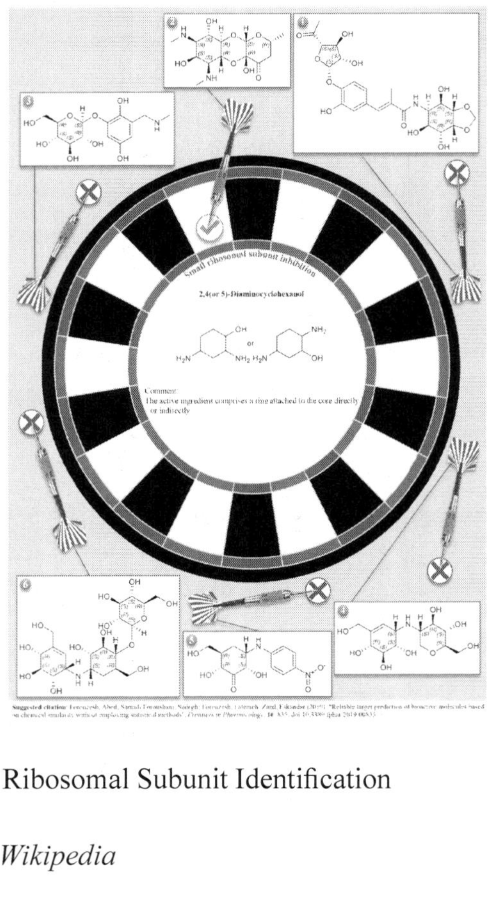

Figure 7.3. Small Ribosomal Subunit Identification

Source: Image by Wikipedia

The mechanisms of tetracyclines and negamycin for the purpose of inhibiting protein synthesis would fairly differ, though they would bind to the overlapping sites in DC. By averting the aa-tRNA to bind to the A site (Jenner L et al., 2013), tetracyclines would inhibit translation. On the contrary, the smaller molecule negamycin would stabilize the binding of tRNA in the A site, hence, translocation would be inhibited and miscoding being incited (Polikanov YS et al., 2014).

The 30S subunit region which consists of h44 and its direct locality would be the target of various groups of antibiotics. Interaction of streptomycin with h44, h27, h18 along with ribosomal protein uS12 would induce conformational modifications of nucleotides A1492 and A1493 (Carter AP et al., 2000, Demirci H et al., 2013).

Aminoglycosides including paromomycin and gentamicin would get bonded at the top of h44, inducing conformational changes of A1492 and A1493 alike the ones led by streptomycin (Carter AP et al., 2000).

Such drugs would support the hyper conformation of A1492 and A1493, thus, increasing the miscoding through stimulation of the near and noncognate binding of the tRNAs in the A site. A different aminoglycoside antibiotic, known as the Hygromycin B, where miscoding would not be stimulated increasing the association of tRNA for the A site and just like negamycin, the tRNA translocation would be inhibited (Borovinskaya MA et al., 2008).

Few of the aminoglycosides, like gentamicin and neomycin would have an additional binding region on the ribosome. On the 50S subunit, they would bind to H69 and ribosome recycling facilitated by RRF might be inhibited (Borovinskaya MA et al., 2007).

7.4.2. Antibiotics Binding with the P and E Sites in the 30S Subunit

The regions occupied by the anticodon loops concerning the peptidyl and deacyl-tRNAs, respectively, at the time of translation would be referred as the P and exit (E) sites in the 30S subunit. The third and final site of binding meant for a tRNA before its dissociation from the ribosome would be the E site.

Inhibitors of translation initiation would generally be the antibiotics through which the P and E sites would be bonded in the 30S subunit. However, the tRNA and mRNA translocation would be affected by a few of them. Such antibiotics comprise of kasugamycin, pactamycin, edeine, tetrapeptide antibiotic GE81112, and amicoumacin A.

Both the P and E sites would be distanced by the binding site of kasugamycin in a place overlapping the mRNA (Schuwirth BS et al., 2006, Schluenzen F et al., 2006).The placement of the drug attached would be incompatible with the established region of the mRNA.

It would further obstruct the formation of the 30S initiation complex. Likewise, the path of mRNA would also be interfered by pactamycin. Further, it would result in preventing the formation of the 30S subunit initiation complex by means of binding the E site of the smaller subunit (Brodersen DE et al., 2000, Polikanov YS et al., 2014).

Edeine, a cationic peptide antibiotic would bind to the small subunit P site and result in forming the base pair between C795 and G693 in the 16S rRNA (Pioletti M et al., 2001). Thus, it would obstruct fMet-tRNAi fMet from bonding to the small subunit. On the other hand, GE81112, a tetrapeptide antibiotic binding to the small subunit P site, would avoid appropriate decoding of the P site by the anticodon-stem loop of fMet-tRNAi fMet being stabilized in an inaccurate conformation (Fabbretti A et al., 2016).

It would result in delay in the protein synthesis steps of initiation due to lower association level for 50S subunit (Fabbretti A et al., 2016). The ribosomal protein uS12 in the subunit 30S would also be considered as a vital constituent of the ribosome even though the ribosome protein uS13 has a significant contribution to the GE81112 binding.

In an attempt to surmount the deficiency of new antibiotics, library screens of natural compounds are supposedly being accomplished for detecting bioactive molecules which could be enhanced more by rational drug design, fragment-based drug discovery as well as computational chemistry.

Thus, identification of the GE82832, a peptide antibiotic made from *Streptosporangium cinnabarinum* (strain GE82832) are considered to be led by such screens. It would include inhibiting tRNA translocation through interaction with the 30S subunit (Brandi L et al., 2006). After further analysis, it was discovered that GE82832 would be associated to an inadequately characterized antibiotic, known as ditromycin, discovered ages ago (Omura S et al., 1977).

7.5. ANTIBIOTICS TARGET THE LARGE SUBUNIT OF THE RIBOSOME

7.5.1. Antibiotics Binding with the Peptidyl Transferase Centre

The binding of most of the antibiotics to the subunit 50S would cluster at the PTC, the location for the formation of the peptide bond, and inside the nascent peptide exit tunnel (NPET). Many among them would inhibit protein synthesis using a conflict with the side chain of amino acid in association with the incoming aa-tRNAs. It would help in binding to the A-site crevice of the ribosome, the wedge-shaped gap formation by the nucleotides bases U2504, A2451 and C2452 belonging to the 23S rRNA (Nissen P et al., 2000).

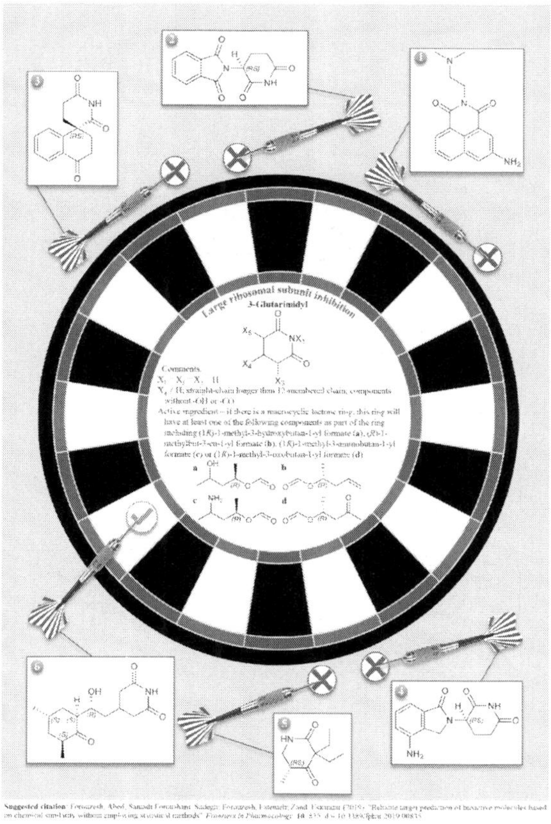

Figure 7.4. Large Ribosomal Subunit Identification

Source: Image by Wikipedia

> Chloramphenicol is considered as one of the antibiotics inhibiting translation in an extensive range of Gram positive and Gram negative bacteria including mitochondria, however, excluding the cytoplasm of the eukaryotic cells. It would be bonded to the A site crevice of the subunit 50S and its nitro benzyl ring would result in the formation of a π-stacking interaction accompanying the base of C2452 of the 23S rRNA.

The binding region of the chloramphenicol aromatic ring overlapping the side chain of amino acid of the incoming aa-tRNA, would adequately prevent it from binding. Chloramphenicol is supposed to be observed as a universal inhibitor facilitating the formation of the peptide bond. It is because every tRNAs would make major interactions with the ribosome in the A site.

However, many experimental analyses, like the various translation inhibition of particular mRNA templates would be left unexplained by the generally accepted chloramphenicol action model (Marks J et al., 2016). Chloramphenicol acting in a content-specified way is supported by growing evidences that prove the dependence of its activity extremely on the nature of precise amino acids belonging to the nascent chain as well as the residual identity that would enter the A site (Marks J et al., 2016).

The antibiotic linezolid which belongs to the oxazolidinone family would be active in controlling all chief pathogenic Gram-positive bacteria. It would be inclusive of methicillin-resistant *Staphylococcus aureus* (MRSA), vancomycin-resistant *Enterococcus faecium*, and penicillin-resistant *Streptococcus pneumonia*. The binding of the linezolid to the ribosome in the A site cleft would be similar to that of the chloramphenicol.

Moreover, on the basis of *H. marismortui*, *S. aureus*, and *Deinococcus radiodurans* 50S subunit crystal structures, the formation of the oxazolidinone ring of linezolid would occur with a π-stacking interaction accompanied by the base of U2504, which would fit appropriately into the A site cleft of the PTC (Ippolito JA et al., 2008, Eyal Z et al., 2015, Wilson DN et al., 2008).

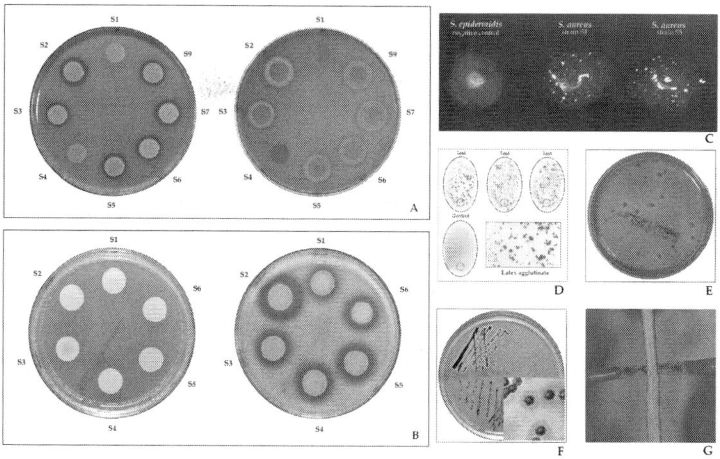

Figure 7.5. *Staphylococcus aureus*

Source: Image by Wikimedia Commons

A semisynthetic derivative, Clindamycin, belonging to lincosamide family of compounds, would be extensively used clinically for treating Gram-positive bacterial infections. The space occupied by the crystal structures of clindamycin along with the *H. marismortui* 50S subunit (Tu D et al., 2005) and the *E.coli* 70S ribosome (Dunkle JA et al., 2010), the pyrrolidinyl propyl group of clindamycin as well as probably other lincosamide, would be same like the aromatic ring of O-methyltyrosine belonging to the CC-puromycin (Hansen JL et al., 2002) and phenylalanine and the remnant of the Phe tRNAPhe binding in the A site (Polikanov YS et al., 2014).

Another group of antibiotics known as Streptogramin A (SA), which would have a binding site spanning not just the A site cleft, but would even impinge into the P site. The simplest SA antibiotic known, Madumycin II, would inhibit the ribosome before the formation of the first cycle of the peptide bond (Osterman IA et al., 2017).

It would let the tRNA to bind to the A and P sites, however, preventing the exact placement of their CCA ends within the PTC. Thus, it results in the forming of peptide bond impossible. Moreover, Madumycin II bond would induce reorganization of nucleotides U2506 and U2585 belonging to the rRNA 23S. It would further result in forming of the U2506•G2583 wobble pair accredited to a catalytically inert state of the PTC (Osterman IA et al., 2017).

Pleuromutilin, like Lefamulin would belong to a group of antibiotics which would actively obstruct the drug-resistant pathogens. By means of forming a bond with the PTC and with interference coordinated by both A and P site tRNAs, they would inhibit protein synthesis.

The crystal structure associated with the *S. aureus* subunit 50S, the tricyclic mutilin core moiety of Lefamulin would result in the formation of hydrophobic stacking interactions together with nucleotides U2504 and C2452 and would attach the drug to the ribosome as well as obstructing the A site. The extended C14 on the tricyclic core of Lefamulin would point within the P site, thus, obstructing the positioning of tRNA at the P and A sites.

7.5.2. Ribosome Inhibition by Antimicrobial Peptides

The cecropins are antibacterial peptides (AMPs) discovered 35 years ago, are considered to have been produced by insects for protecting them against pathogen infections (Steiner H et al., 1981). Presently known AMPs have been broadly expressed in both plants and animals. Being powerful therapeutic agents, one of the sub-categories of AMPs known as the proline-rich antimicrobial peptides (PrAMPs) would be expressed in both plants and animals.

Due to their extensive distribution and their ability to kill bacteria through mechanisms that would not result in disrupting the cell-membrane, they have gained a lot of popularity (OtvosL Jr., 2000). Few of the PrAMPs, like oncocin, apidaecin, drosocin, pyrrhocoricin, metalnikowin, and bactenecin which would be interacting with the micromolar affinity with the substrate-binding area belonging to the chaperone DnaK and ultimately would lead to protein misfolding and aggregation as well as bacterial death eventually (Otvos L Jr. et al., 2000, Otvos L Jr., 2005).

However, *E. coli* strains lacking in DnaK would remain vulnerable to PrAMPs. It is an indication that there would be other targets inside the PrAMPs cell (Krizsan A et al., 2014). The secret of the primary cellular target of PrAMPs derived from insects would have remained unexplained till the time it was found that the 70S ribosome would be bonded by oncocin and apidaecin showing nanomolar affinities (Krizsan A et al., 2014).

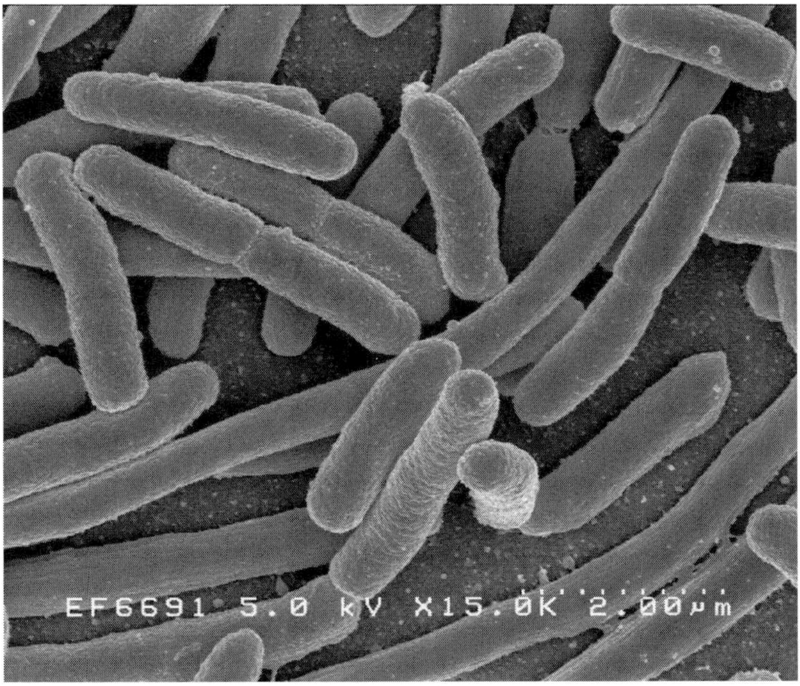

Figure 7.6. *Escherichia coli*

Source: Image by Wikipedia

Thus, the results would help to comprehend the reason for DnaK null mutant *E.coli* strains staying vulnerable to apidaecin and oncocin. Similarly, it was stated that bactenecin 7 of the cathelicidin family would be inhibiting protein synthesis by interacting with the 70S ribosome (Mardirossian M et al., 2014).

Biochemical and structural studies related to various ribosome complexes in associated with oncocin, metalnikowin, pyrrhocoricin, and bactenecin 7- it has been discovered that PrAMPs would be binding inside the exit tunnel of ribosome following a reversal orientation in contrast to that of a nascent polypeptide chain (Graf M et al., 2017, Seefeldt AC et al., 2016, Gagnon MG et al., 2016, Seefeldt AC et al., 2015, Roy RN et al., 2015).

The area where PrAMPs bond in the peptide exit tunnel and their overlapping with the PTC would result in letting the protein synthesis initiation to occur. However, it would act in preventing the transition to the elongation stage of translation to be interfered by the bonding of the aa-tRNA in the A site of the subunit 50S.

Although in the majority of the lately studied PrAMPs translation initiation would be inhibited, yet, apidaecin is supposed to have exhibited translation termination inhibition (Florin T et al., 2017). The ribosomal RF1 or RF2 would be trapped by apidaecin corresponding to the nascent peptide chain being released.

Reconstruction of the *E.coli* 70S ribosome through cryo-EM compounded with RF1 and Api137 (Florin T et al., 2017), considered to be extra stable 18-amino acid derived from the natural apidaecin 1b (Berthold N et al., 2013), would tell that Api137 too would be binding into the ribosomal peptide exit tunnel. Further, the C-terminal amino acids pertaining to it would be reacting with nucleotides 23S rRNA in the PTC as well as with the RF1 GGQ motif.

In contrast to other PrAMPs, wherein, structures would be present, the Api137 orientation would match with that of a nascent polypeptide chain. Although numerous inhibitors, like blasticidin S, bactobolin A, and negamycin, are considered to have been reported for interference with translation termination (Svidritskiy E et al., 2013, Amunts A et al., 2015, Uehara Y et al., 1976), yet, such antibiotics would even be hampering other steps of protein synthesis.

Hence, apidaecin would be the first inhibitor recognized distinctly aiming at the termination step of translation. Such PrAMPs would interact with the ribosomal exit tunnel and since they would overlap the binding regions of various other drugs would give scope to such peptides for developing into novel therapeutics.

7.6. ANTIBIOTIC RESISTANCE MECHANISM

7.6.1. General Mechanisms of Antibiotic Resistance

The antibiotic resistance development would frequently follow their introduction rapidly within the clinic. There would be no room for exceptions for the antibiotics targeting the ribosome. Since antibiotics produced naturally are supposed to have been existing in the environment for a countless number of years, antibiotic resistance nevertheless would be considered to be an ancient feature. Moreover, the resistance was developed among environmental and pathogenic bacteria much ahead of modern antibiotic usage for curing infections (D'Costa VM et al., 2011).

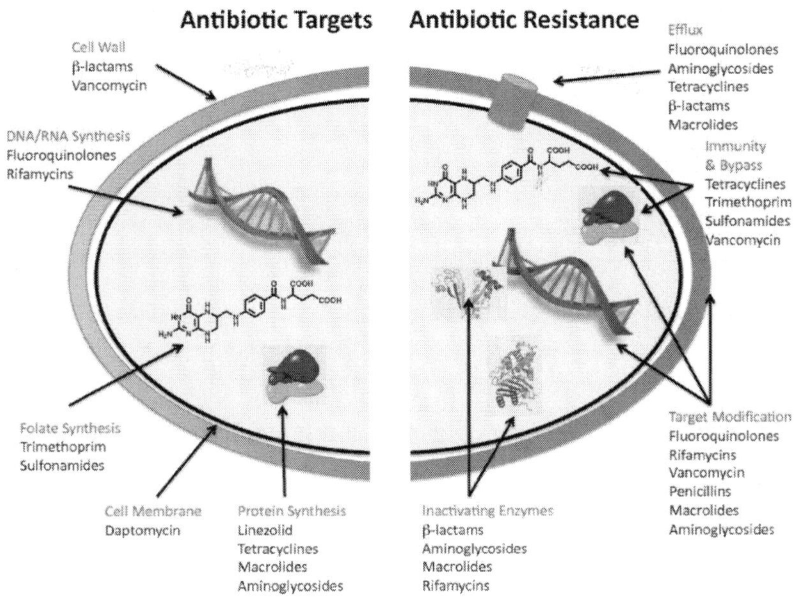

Figure 7.7. Antibiotic Resistance Mechanisms

Source: Image by Wikimedia Commons

Antibiotic resistance would be associated with their production inherently. It is because microorganisms producing antibiotics are believed to be efficient enough to safeguard themselves from the action of the drugs they would make. Such fundamental mechanisms of resistance prevailing in nature would set the foundation aimed at the acquired resistance in pathogens subjugated to a particular pressure produced by the usage of antibiotics by humans.

Pathogens would be able to gain resistance genes with the help of horizontal gene transfer (HGT), on mobile components, for example, plasmids, transposons, or integrons, and they might have evolved from the environment. Together with HGT, spontaneous point mutation or chromosomal mutation could be other ways for pathogens for gaining resistance.

There would be three different ways for bacteria to become antibiotic resistant:

- by managing to maintain the intracellular concentration of the drug minimum

- by chemically making the drug inactive
- by changing its target to restrict it from interacting with the drug anymore

Being a front line defender, antibiotic concentrations could be regulated by bacterial cells either by limiting their flow of influx or by improving their flow of efflux from the cell. Few bacteria would be essentially resistant towards specific antibiotics due to reduced uptake of such drugs.

For example, Gram-negative bacteria would show resistance towards macrolides. It is because the hydrophobic nature of macrolides would be preventing them to penetrate from their outer membrane.

Anaerobic bacteria are considered to have acquired natural resistance for aminoglycosides. It is because aminoglycosides transportation into the cell would be driven by means of the energy that is produced during oxidative metabolism, something that would be impossible for such bacteria to perform. Additionally, resistance could be developed by bacteria augmenting the efflux of a specific drug, thus, minimising its concentration to a nontoxic level within the cell.

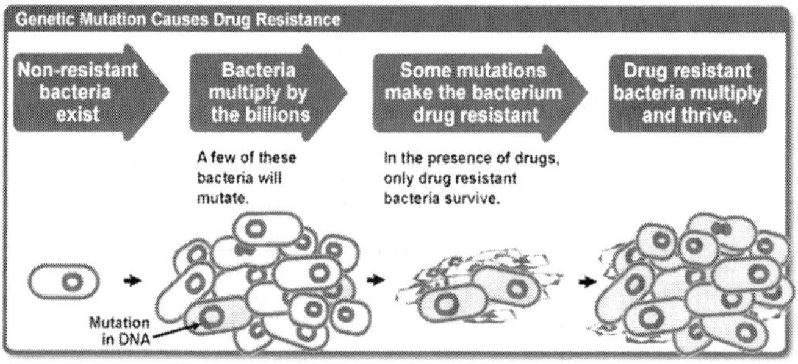

Figure 7.8. Genetic Mutation by Drug Resistance

Source: Image by Flickr

It could be gained by acquiring encoded genes pertaining to drug specific efflux pumps or chromosomal mutations which would result in over-expressing a pre-existent efflux pump, like the ToIC multidrug efflux pump in *E. coli* (Li XZ et al., 2015).

The upregulation of the efflux pumps would be conferring resistance in drugs which would be recognized for nearly all antibiotics targeting ribosome. In the case of a few antibiotics, drug resistance was supposed to have been indicated through the improved activity of porins.

The antibiotics could be made inactive either by enzymatic modification (e.g., aminoglycosides) or by means of degradation (e.g., tetracyclines) when it is inside the cell. Likewise, such enzymes encoded by genes could be found either by HGT or mutation.

Altering the target would be the third common principle to acquire drug resistance. Binding of antibiotics through physical alteration of their target would be reduced by mutations at or close to the binding regions. The binding regions for commonest antibiotics targeting ribosomes would be mostly made of rRNA.

Thus, resistance could be conferred by rRNAs encoded genes mutations. However, the maximum number of bacteria would contain numerous replicas of rRNA operons. Moreover, provided the same mutation occurring at the same time on every replica, the resistance would be manifested on its own.

Hence, in the case of worst clinical pathogens, antibiotic resistance from rRNA mutation would be occasional. Yet, pathogens like *Mycoplasma pneumoniae*, *Mycobacterium tuberculosis* and other species of the bacteria with only one or two rRNA operons, would encounter such mutations (Vester B et al., 2001).

In few of the odd cases, antibiotics and ribosomal proteins would bond directly or might find ribosomal protein from the location of their binding sites. Resistance would develop more often in such antibiotics through ribosomal protein mutation; whereas, a single replica in the genome would generally contain the genes. Together with mutation, the enzymatic modification could also change the binding sites of antibiotics, which would result in all the rRNA copies being affected in the cell, thus, developing strong resistance.

Though, for any given antibiotic, there would be varied mechanisms of resistance, yet, one mechanism would frequently be prevailing for several pathogens. Following would be some of the vital mechanisms of resistance occurring from the clinical usage of certain vital antibiotics targeting ribosome:

7.6.2. Mechanisms of Resistance to Aminoglycosides

Cellular enzymes modifying drugs would be a popular mechanism of resistance in case of aminoglycosides. Due to amide and hydroxyl groups being exposed, aminoglycosides would be susceptible to enzymatic modification. More than 300 enzymes, which could be classified into three groups, would include acetyltransferases, nucleotidyltransferases, or phosphotransferases.

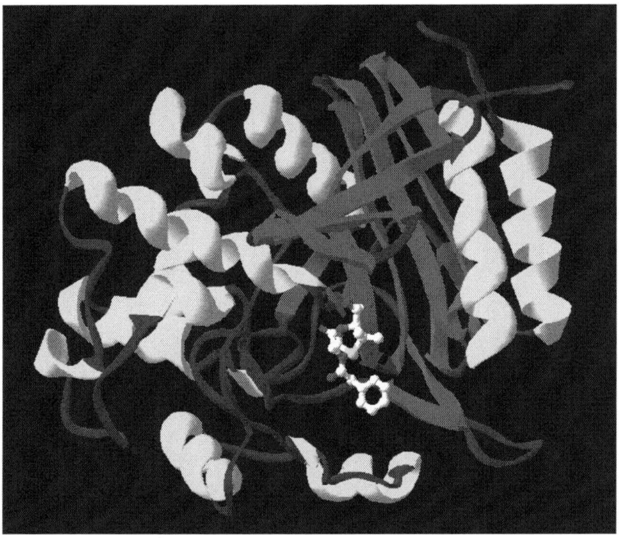

Figure 7.9. Enzyme Inhibitor

Source: Image by Wikipedia

Such enzymes would be modifying the aminoglycosides found in the amino groups or hydroxyl groups and would be the source behind losing their ribosome binding activity (Garneau et al., 2016). The said enzymes which would have probably evolved from bacteria producing aminoglycoside, would presently exist in transmissible elements presenting a major clinical threat. It is due to the fact that they would be conferring resistance to a wide array of aminoglycosides.

Ribosomal mutations in pathogens which would contain one or two replicas of rRNA operons would be able to confer aminoglycosides resistance. For example, in h44 of the 16S rRNA, the A1408 mutation is considered to have been found in resistant *M. tuberculosis* clinical isolates (Maus CE et al., 2005).

Remarkably, in human cytosolic ribosomes, normally G1408 would exist and would explain the reason behind aminoglycosides being inactive against human ribosomes justifying their bacterial specificity. Unlike the majority of aminoglycosides binding h44, streptomycin would be interacting with the G530 loop (h18) along with uS12 ribosomal protein.

Hence, the resistance mechanism which would occur often to streptomycin in *M. tuberculosis* would be one among the two lysine residues K43 or K88 mutation, in which both would contact ribosome bound streptomycin (Honore N et al., 1994).

7.6.3. Mechanisms of Resistance to Tetracyclines

Acquiring mobile genetic components containing specific tetracycline resistant genes which would be encoding for efflux pumps, ribosome protection proteins (RPPs) and enzymes degrading drug, would lead to tetracycline resistance.

Specific efflux pumps pertaining to tetracycline would be highly beneficial against tetracyclines of the first generation. They would provide moderate resistance to the tetracyclines of the second-generation, like the doxycycline and minocycline, as well as slight resistance to the tetracycline of the third generation (also known as glycylcyclines), like the tigecycline (Grossman TH., 2016).

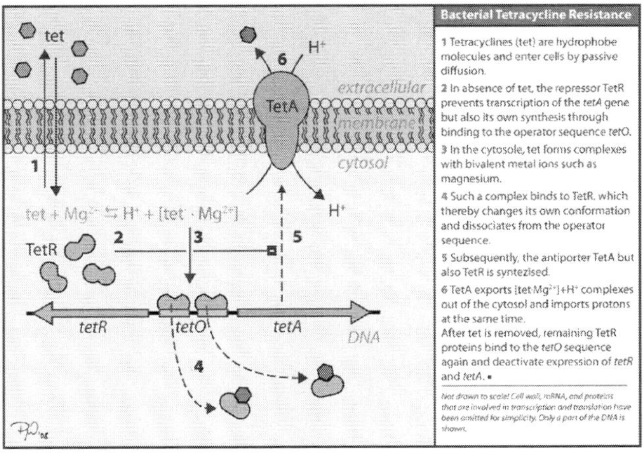

Figure 7.10. Tetracycline Resistance

Source: Image by Wikimedia Commons

Another way of resisting tetracycline inhibitions would be by the usage of RPPs. Representations of RPPs by TetO and TetM would be EF-G paralogs (Connell SR et al., 2003). The tetracycline-stalled ribosome would be bonded with them, causing tetracycline dissociation from the ribosome.

The TetM reconstructed by cryo-EM, bounded to the *E. coli* 70S would be showing Pro509 at the vertex of the domain IV being overlapped with the tetracycline binding position and would be signifying a dislodging ribosome-bound tetracycline mechanism (Arenz S et al., 2015).

Interestingly, the tigecycline binding to the ribosome would not be affected by TetM, although anticipation of more extreme conflict between TetM and tigecycline would be made. Lately, the ABC-F subfamily members belonging to the ATP binding cassette proteins were considered to have been seen protecting the ribosome from antibiotics inhibition that would bind close to the PTC, like macrolides, streptogramins, lincosamides and oxazolidinones (Sharkey LK et al., 2016).

Enzymatic degradation is supposed to be a tetracycline resistance mechanism which would be uncommon in the clinic. Tetracycline would be modified at position C11a by a flavin-dependent monooxygenase TetX which would be encoded by TetX gene. It would result in generating 11a-hydroxyl tetracycline, undergoing an impulsive nonenzymatic breakdown.

For years, TetX was supposed to be the only tetracycline degrading enzyme being biochemically characterized as well as considered to have been found in some human pathogens (Leski TA et al., 2013). However, recently, destructases, belonging to a family of extensive TetX- related genes, have been discovered (Forsberg KJ eta al., 2015).

Such genetic products would apply several mechanisms for inactivating the tetracycline, but would exclude tigecycline (Park J et al., 2017). The destructases family existing would trigger an alarm of trouble. Once dispersed within human pathogens and mutated, the gene would be aggravating the problems instigated due to tetracycline resistance as well as pose a potential risk to the tigecycline clinical usage (Linkevicius M et al., 2016).

7.6.4. Mechanisms of Resistance to Macrolides, Lincosamide, Streptogramins B

In few antibiotics, they would share binding positions that would overlap in the NPET on the subunit 50S, like macrolides, lincosamides, streptogramins B and ketolides (collectively they would be called MLSBK). Hence, it is considered that development of resistance against one of the above mentioned drug would generally get extended to the other members of the MLSBK. Mono or dimethylation of A2058 of the 23rRNA would be regarded as a common method of MLSBK antibiotics resistance. It would be situated at the upper part of the NPET.

Such modifications would act as a preventive measure towards the binding of every MLSBK antibiotics to the ribosome. A group of plasmid-encoded resistance enzymes, known as erythromycin resistance methyltransferases is supposed to be catalysing such type of modification in the ribosome. Since the fitness of the cells would be decreased by A2058 methylation (Gupta P et al., 2013), hence, through a well-studied translation attenuation mechanism, erythromycin and additional macrolides would be usually induced in the genes of erythromycin resistance methyltransferase (Ramu H et al., 2009).

A separate methyltransferase, Cfr, would target A2503 of the 23rRNA and the resultant activity would form a dimethylated m2m8A2503 residue. Further, a wide range of antibiotics, like phenicol, lincosamides, oxazolidinones, pleuromutilin and streptogramins A, would be conferred by it (Polikanov YS et al., 2015, Giessing AM et al., 2009).

Interestingly, the chromosome belonging to the MRSA strain CM05 would have a single operon containing ermB and cfr genes. The synergistical modification of both A2058 and A2503 would improve the resistance to ring macrolides of 16 members (Smith LK et al., 2008).

A2058 or A2059 mutation are supposed to be present in pathogens with one or two rRNA operons, the same that would be conferring resistance to MLSBK antibiotics (Fyfe C et al., 2016). Additionally, ribosomal protein mutations uL4 and uL22 could be reducing the vulnerability towards one or more MLSBK antibiotics. Such proteins would contain extended loops forming the peptide exit tunnel wall, which would be close to the binding region of MLSBK antibiotics.

Though there would be no direct contact of the ribosome bound drug with such loops, yet, their mutations could result in triggering conformational transformations related to nearby rRNA propagating into the MSBK antibiotics binding sites (Wekselman I et al., 2017).

7.7. CONCLUSION

With the emergence of antibiotic usage in the 1940s, a significant improvement in infectious disease outcomes was noticed. Over the past seven decades, it is supposed to have acted as a preventive measure in treating numerous complications and deadly diseases. However, excessive usage of antibiotics in contemporary medicine, livestock and agriculture would have resulted in the development of drug-resistant strains. It is believed to have restricted the efficiency of several drugs clinically and incited the increasing global threat of the outbreak caused by antibiotic resistant pathogens (Lowell AN et al., 2015).

As per studies, it has been observed that resistance towards a single antibiotic would result in often extremely limiting the clinical effectiveness of all the structurally similar drugs pertaining to the same group. Few of the examples would be beta-lactamase expression towards penicillin resistance, export proteins expression in tetracyclines resistance and ribosomal protein resistance in aminoglycosides resistance, like streptomycin (Wilcox SK et al., 2001).

The repertories of antibiotics would be extremely constrained by the restricted count of structural classes of antibiotic including an inadequate number of molecular targets, such as enzymes of the bacterial cell wall and the ribosomes of the prokaryotes (Polikanov YS. Et al., 2014). Thus, one of the approaches that would help in combatting the perseverance of clinical drug resistance would be to find antibiotics of new structural or target-based groups and which would act according to novel mechanisms.

REFERENCES

1. Chellat, M., Raguž, L. and Riedl, R., 2016. Targeting Antibiotic Resistance. *Angewandte Chemie International Edition*, [online] 55(23), pp.6600-6626. Available at: <https://www.ncbi.nlm.nih.gov/pmc/articles/PMC5071768/> [Accessed 9 May 2020].
2. Lin, J., Zhou, D., Steitz, T., Polikanov, Y. and Gagnon, M., 2018. Ribosome-Targeting Antibiotics: Modes of Action, Mechanisms of Resistance, and Implications for Drug Design. *Annual Review of Biochemistry*, [online] 87(1), pp.451-478. Available at: <https://www.annualreviews.org/doi/full/10.1146/annurev-biochem-062917-011942> [Accessed 9 May 2020].
3. Lin, J., Zhou, D., Steitz, T., Polikanov, Y. and Gagnon, M., 2018. Ribosome-Targeting Antibiotics: Modes of Action, Mechanisms of Resistance, and Implications for Drug Design. *Annual Review of Biochemistry*, [online] 87(1), pp.451-478. Available at: <https://www.researchgate.net/publication/323976028_Ribosome-Targeting_Antibiotics_Modes_of_Action_Mechanisms_of_Resistance_and_Implications_for_Drug_Design> [Accessed 9 May 2020].
4. Lowy, F., n.d. *Protein Synthesis Inhibitors*. [ebook] p.3. Available at: <http://www.columbia.edu/itc/hs/medical/pathophys/id/2009/proteinNotes.pdf> [Accessed 9 May 2020].
5. Novac, O., 2004. Inhibitors of protein synthesis identified by a high throughput multiplexed translation screen. *Nucleic Acids Research*, [online] 32(3), pp.902-915. Available at: <https://www.ncbi.nlm.nih.gov/pmc/articles/PMC373382/> [Accessed 9 May 2020].
6. Orelle, C., Carlson, S., Kaushal, B., Almutairi, M., Liu, H., Ochabowicz, A., Quan, S., Pham, V., Squires, C., Murphy, B. and Mankin, A., 2013. Tools for Characterizing Bacterial Protein Synthesis Inhibitors. *Antimicrobial Agents and Chemotherapy*, [online] 57(12), pp.5994-6004. Available at: <https://aac.asm.org/content/57/12/5994> [Accessed 9 May 2020].
7. Tae, H. and Thakral, D., 2017. *Focus: Drug Development: Discovery Of A Structurally Unique Small Molecule That Inhibits Protein Synthesis*. [online] PubMed Central (PMC). Available at: <https://www.ncbi.nlm.nih.gov/pmc/articles/PMC5369043/> [Accessed 9 May 2020].
8. VAZQUEZ, D., 1974. *INHIBITORS OF PROTEIN SYNTHESIS*. [ebook] p.2. Available at: <https://core.ac.uk/download/pdf/82687868.

pdf> [Accessed 9 May 2020].

9. Witzky, A., Tollerson, R. and Ibba, M., 2019. Translational control of antibiotic resistance. *Open Biology*, [online] 9(7), p.190051. Available at: <https://www.ncbi.nlm.nih.gov/pmc/articles/PMC6685928/> [Accessed 9 May 2020].
10. Amunts A et al., 2015
11. Arenz S et al., 2015
12. Berthold N et al., 2013
13. Borovinskaya MA et al., 2008
14. Brandi L et al., 2006
15. Brock, 1961,
16. Brodersen DE et al., 2000
17. Carter A.P. et al. 2000
18. Charani E et al., 2019
19. Cochella L et al., 2004
20. Connell SR et al., 2003
21. Cundliffe, 1972
22. D'Costa VM et al., 2011
23. Demirci H et al., 2013
24. Dunkle JA et al., 2010
25. Durand GA et al., 2018
26. Ero R et al., 2019
27. Ettayebi M et al., 1985
28. Eyal Z et al., 2015
29. Fabbretti A et al., 2016
30. Florin T et al., 2017
31. Forsberg KJ et al., 2015
32. Francklyn CS et al., 2019
33. Fyfe C et al., 2016
34. Gagnon MG et al., 2016
35. Gale, 1963
36. Garneau et al., 2016
37. Giessing AM et al., 2009

38. Gold L. et al., 1995
39. Gottlieb et al., 1967
40. Graf M et al., 2017
41. Graff J.R. et al., 2003
42. Gregory ST et al., 2009
43. Grossman TH., 2016
44. Gupta P et al., 2013
45. Hahn et al., 1951
46. Hansen JL et al., 2002
47. Harvey I. et al., 2002
48. Hershey J.W. et al., 2000
49. Ho JM et al., 2018
50. Honore N et al., 1994
51. Hummel H et al., 1987
52. Ippolito JA et al., 2008
53. Jenner L et al., 2013
54. Krizsan A et al., 2014
55. Lambert PA., 2002
56. Leski TA et al., 2013
57. Levy SB et al., 2004
58. Li XZ et al., 2015
59. Lin J et al., 2018
60. Linkevicius M et al., 2016
61. Mandal M. et al., 2003
62. Mardirossian M et al., 2014
63. Marks J et al., 2016
64. Mironov A.S. et al., 2002
65. Monshupanee T et al., 2008
66. Muiioz et al.,1972
67. Nahvi A. et al., 2002
68. Newman DJ et al., 2012
69. Newton et al., 1966
70. Nguyen F et al., 2014

71. Nirenberg, M.W. and Matthaei, J.H., 1961
72. Nissen P et al., 2000
73. Ogle JM et al., 2001
74. Omura S et al., 1977
75. Osterman IA et al., 2017
76. OtvosL Jr., 2000
77. Park J et al., 2017
78. Peach KC et al., 2013
79. Pestka S., 1977
80. Pioletti M. et al., 2001
81. Poehlsgaard J et al., 2005
82. Polikanov YS et al., 2014
83. Ramu H et al., 2009
84. Roy RN et al., 2015
85. Ruggero D. et al., 2003
86. Sander P et al., 1996
87. Schulze CJ et al., 2013
88. Schuwirth BS et al., 2006
89. Seefeldt AC et al., 2016
90. Sharkey LK et al., 2016
91. Smith LK et al., 2008
92. Springer B et al., 2001
93. Svidritskiy E et al., 2013
94. Tenson T et al., 2006
95. Tu D et al., 2005
96. Uehara Y et al., 1976
97. Vazquez D., 1979
98. Ventola CL., 2015
99. Vester B et al., 2001
100. Watkins S.J et al., 2002
101. Weisblum et al., 1968
102. Wekselman I et al., 2017
103. Werstuck G. et al., 1998

104. Wilcox SK et al., 2001
105. Wilson DN. 2014
106. Wong WR et al., 2012

8

Evolution of the Translation Apparatus

CONTENTS

This chapter is about how the translation apparatus has evolved over the time. The first half of the chapter discusses the meaning and the origins of the translation and the timing information. The subsequent topics include the insights to the ribosomal history from tRNA structure: basic theories of the code nature, origin, and evolution. The mid half of the chapter has everything about the origin of tRNA and sequence-based analysis of AARSs. The chapter ends with the topics explained such as an RNA-FIRST view of the origin of life, Abiotic Synthesis of Polynucleotides, Nonenzymatic replication of RNA, the first RNA replicase etc.

8.1. INTRODUCTION

The most conserved cellular proteins are found in the translation apparatus. It also includes the RNAs which provides the evidence for all forms of life with the shared ancestry. The tRNAs and their cognate amino-acyl tRNA synthetases, along with the ribosomal RNAs and highly conserved ribosomal proteins help in the formation of the peptide with that of amino acid building blocks.

Some conserved protein factors are found and observed at each step of translation. The steps involved in the translation process are initiation, elongation, and release. These conserved protein factors are required for the actual ribosomal decoding of mRNA. These are the GTPases primarily which ensures that the decoding of the genetic code is done accurately by acting as proofreading switches.

The RNA world is seen as the culmination of chemical evolution by the current theories based on the origin of the life. The subject of the debate is the progenotes that is responsible for populating its biochemical complexity and the RNA world.

The workers have a common point of agreement in the field that for the chiral synthesis of defined sequence peptides that the discovery of machinery would have made the way for the modern world of protein. Initially, with the discovery of RNA replicase was a catalytic RNA or an early peptide product. Now the stage would be set for those having both the features in one enclosure which would help in the development of populations of progenotes.

The advanced progenotes like this would be considered as the first entities that are capable of having the genetic couple among the transcription, replication, and translation. At one stage, by ribosomes, the modern day tmRNA works as tRNA, while on the other stages, in the translation process, it serves as mRNA.

The above discussed variations of RNA might be the types of entities present in the late RNA World. The DNA can be used as a storage medium that would store the genetic information. It would help in the finalization of the transition into the modern world from the progenotic world. This would be transferred into the living systems of the modern world.

According to the origins, as discussed above, the beginning of the cellular protein synthesis machinery into the modern world started before the emergence of the last common ancestor of modern life. The question arises here is that Is this enough to know about the origins? It is rightly said that the genes that the descendants hold are possibly the genes of the ancestors. However, there are chances of gaining or losing the polyphyletic gene.

This way there were several steps or attempts were undertaken to outline the last common ancestors. With relation to the translation machinery, the properties that have similar things is rather extensive. This attempt was challenged by a biologist, Doolittle. Doolittle said that the results might be confusing when the possibility is that of extensive lateral gene transfer. However, this view was defended by Glansdorff, who argued that there is overemphasis given to the significance of lateral gene transfer.

There is another approach which is very different from the above discussed. It is related to the examination of the information processing systems in toto. The information processes in the three major lineages were compared when all three Domains of the complete genome data became available for the first time.

As per the above discussions, there was a conclusion which says that the DNA replication is unique to the lineage of the (EU) bacterial machinery. The examination of each of these issues was done and it was reassessed by examining the essential proteins. The conclusions were reaffirming. The transcription case showed that all the three lineages are having the archaeal and eukaryal versions as the core components are similar.

8.2. LANDMARK ADVANCES IN THE DEVELOPMENT OF CELL-FREE PROTEIN SYNTHESIS SYSTEMS

Because the preparation processes did not remove mRNA that is messenger RNA, the cell extracts (which contain early cell-free systems) further synthesized native proteins most of the time from the endogenous mRNA. These extracts were considered to be essentially supernatants and they were obtained by centrifugation at high speed of lysates from disrupted cells. An example can include 30,000g (this is what is referred to as S30 extract) (Lamborg and Zamecnik, 1960; Matthaei and Nirenberg, 1961; Schweet et al., 1958; Zamecnik et al., 1948).

Later, Nirenberg and Matthaei incubated the *E. coli* S30 extract which was done in the presence of 20 amino acids and ATP regeneration components (Nirenberg and Matthaei, 1961) to direct the protein synthesis machinery for translating the exogenous mRNA.

This type of "incubated S30" further improved the amount of the protein which was synthesized from exogenous mRNA. This is done possibly by freeing ribosomes from getting attached to endogenous mRNA (which is referred to as ribosome run-off) and by allowing endogenous mRNA that is to be degraded by ribonucleases in the extract.

In a similar way, Pelham and Jackson added a micrococcal ribonuclease that is Ca^{2+}-dependent while preparing the rabbit reticulocyte extract (that is lysate) to degrade endogenous mRNA (Pelham and Jackson, 1976). Subsequently, the micrococcal ribonuclease can be inactivated by EGTA which is done to prevent its digestion of exogenous mRNA.

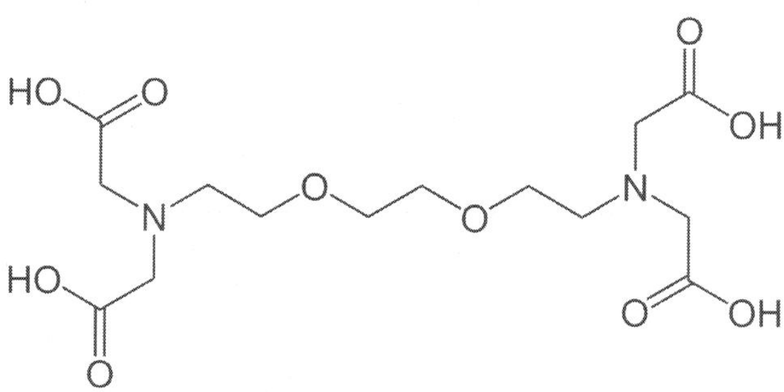

Figure 8.1. EGTA

Source: Image by Wikimedia Commons

There is no translation of a significant amount of endogenous proteins that takes place from the wheat germ extract. This is the reason why there is no need for incubation or nuclease-treatment (Roberts and Paterson, 1973).

There was the next landmark while improving the cell-free protein synthesis. This was the introduction of coupled transcription-translation systems. These were the systems in which the DNA acted as the template rather than mRNA (Chen and Zubay, 1983; Zubay, 1973). *E.coli* RNA polymerase was used by Zubay and colleagues in the S30 extracts for transcribing the DNA into mRNA for protein synthesis, and majorly for performing the study on the gene regulation *in vitro*.

Other researchers later used stronger phage promoters and the more efficient phage RNA polymerases (T7 or SP6) for coupled transcription-translation from the DNA template to obtain higher amounts of mRNA (Craig et al., 1992; Krieg and Melton, 1987; Nevin and Pratt, 1991).

The coupled transcription-translation is done based on the phage promoters and RNA polymerases. This translation has been adopted by most cell-free systems since then. Along with this while using the DNA template which was more stable and avoiding a separate *in vitro* transcription step, the coupled transcription and translation in the system that is cell-free result in a higher protein synthesis yield most of the time.

This is because the newly synthesized mRNA is immediately translated into protein and this can further reduce the adverse effects of mRNA degradation or inhibitory structures.

The conduction of the cell-free protein synthesis is commonly done in batch format. Nevertheless, the protein synthesis reactions stop after a short period of time most of the time and there are several cases the reactions stop in only 1 hr or so. This is because of the depletion of ATP and GTP, degradation of mRNA and collection of by-products for inhibiting concentrations.

Normally, the batch cell-free protein synthesis tends to yield μg ml−1 of protein. The introduction of the concept of the continuous cell – free protein synthesis in both *E.coli* as well as eukaryotic cell-free systems was done by Spirin and colleagues (Spirin et al., 1988; Ryabova et al., 1998; Shirokov et al., 2007).

The continuous cell-free protein synthesis tends to circumvent some issues of the batch format and this is done by using a flow chamber or dialysis membrane. This flow chamber or dialysis membrane is used to maintain the levels of small molecules energy sources as well as amino acids and helps in removing inhibitory by-products such as phosphate, at the same time. This further resulted in a continuous cell-free protein synthesis reaction which can last more than 10 hours most of the time and yields mg ml−1 of protein.

Because of the continuous improvement that is done by academic and industrial researchers, the improvements have been done by cell extracts in cell-free systems significantly. Nirenberg and Matthaei were the first described the "incubated S30" (Nirenberg and Matthaei, 1961). The development of the robust protocols was done by Zubay, Pratt and colleagues to make the *E. coli* extract which has become the standards (Pratt, 1984; Zubay, 1973).

There were some other important improvements which further included the use of *E. coli* strains that lacked RNase I (rna) (MRE600) and Exonuclease V (A19) to stabilize RNA and linear DNA templates, respectively (Gesteland, 1966; Yang et al., 1980).

, Swartz and colleagues pioneered the use of genome engineered *E. coli* strains. This is done to stabilize amino acids during *in vitro* reactions (Michel-Reydellet et al., 2004), and during the *in vitro* activation of endogenous metabolic pathways to sustain high - yield protein synthesis (Jewett et al., 2008; Jewett and Swartz, 2004a), as well as the systematic optimization of *E. coli* growth conditions and extraction procedures for the reduction of the production cost (Kim et al., 2006; Liu et al., 2005; Zawada and Swartz, 2005; Zawada and Swartz, 2006).

Since the time of their initial development, there was no change in the protocols to make the "classic" eukaryotic extracts, rabbit reticulocyte lysate (Jackson and Hunt, 1983) and wheat germ (Anderson et al., 1983; Erickson and Blobel, 1983). Moreover, the protein synthesis yields from these eukaryotic extracts are generally lower in comparison to the *E. coli* extract. In the recent period, many researchers developed the improved protocols to make highly active extracts from wheat germ and mammalian cells.

The remnants of endosperm on the surface of wheat germ, as per the study done by Endo and colleagues, were the sources of the ribosome-inactivating proteins as well as other translation inhibitors. These were considered to be responsible for the poor stability as well as the low yield of the conventional wheat germ cell-free system (Madin et al., 2000). Endo and colleagues further developed a high – yield long-lasting wheat germ cell-free system which was done by extensively washing isolated wheat germ (Madin et al., 2000; Sawasaki et al., 2007).

Till the time now, the extracts of mammalian cells except the rabbit reticulocyte lysate have not been widely used for cell-free protein synthesis majorly because of the low yields. At RIKEN, Imataka and colleagues found that in the HeLa cell extract, the lack of functional translation initiation factors was a major limiting factor (Mikami et al., 2006b).

The development of the significantly improved mammalian cell-free systems were done by the Imataka group and this was done by supplementing with exogenous translation initiation factors (Mikami et al., 2010; Mikami et al., 2006a; Mikami et al., 2006b). Moreover, this was done by effectively taking over the rabbit reticulocyte lysate system.

Ultimately, the Ueda group reconstituted the minimal protein translational machinery from *E. coli*. This was done by completely circumventing the issues of inhibitory factors in cell extracts (Shimizu et al., 2001). The PURE system that is Protein synthesis Using Recombinant Elements was created by the Ueda group and this system consists of wholly purified components along with minimal nucleases, proteases, and other enzymatic activities which are detrimental or unrelated to protein translation.

Although other laboratories had done many attempts to reconstitute the protein synthesis along with purified components (Ganoza et al., 1985; Kung et al., 1977), the PURE system was considered to be the first robust cell-free system which contained a complete set of purified translation factors and the aminoacyl-tRNA synthetases (Shimizu et al., 2001; Shimizu et al., 2005).

Keeping the same approach in mind, the laboratory reconstituted the minimal protein translational machinery from *Thermus thermophilus*. Furthermore, it demonstrated cell-free protein synthesis at temperatures up to 65°C (Zhou et al., 2012).

Swartz and colleagues pioneered chemical engineering approaches for making the *E. coli* extract, to make cell-free protein synthesis by using a protein factory and not just an analytic tool. Their main aim was to manufacture therapeutic proteins in the cell-free system at a liter or any scale (Carlson et al., 2012).

The whole processes of cell growth, extract preparation, protein synthesis and downstream purification were considered for the cost analyses as well as process engineering. The "black-box" nature of the *E. coli* extract was then acknowledged as the sets of biochemical reactions amenable to pathway engineering.Among much key advancement, one of them was the creation of the "Cytomim" system. This system can activate central metabolism and oxidative phosphorylation in the *E. coli* extract, furthermore, allowing the use of inexpensive energy substrates, such as glutamate or glucose, and also, continuous protein synthesis and that too for more than 6 hours in the batch format (Jewett et al., 2008; Jewett and Swartz, 2004a; Jewett and Swartz, 2004b).The Cytomim system was considered to be the basis for further improvements. These improvements resulted in the cell-free synthesis of recombinant human granulocyte-macrophage colony-stimulating factor in a 100-liter industrial stir tank (Jewett et al., 2008; Zawada et al., 2011).

8.3. DEVELOPMENT OF THE PROTEIN SYNTHESIS FIELD

Considering this view, it can be said that the ribosome and other components of the protein synthesis machinery contain a relatively stable decoding as well as synthetic apparatus which is programmed by unstable mRNA. Soon, this was confirmed in bacteria as well as bacterial cell-free systems (Brenner et al. 1961; Gros et al. 1961).

There was a discovery that the poly (U) can provide directions to the synthesis of polyphenylalanine *in vitro* (Nirenberg and Matthaei 1961). This discovery was a transformative event while spearheading the elucidation of the genetic code by the period of the mid – 1960s. There was a wobble hypothesis and this hypothesis rationalizes the features of the redundancy of the code and its decoding by tRNAs. This wobble hypothesis was published by Crick (1966). Considering the higher cells, it was further recognized that there was the existence of a class of rapidly labeled RNA was heterogeneous in size and having distinct chromatographic properties.

It has an essential feature and that is informational intermediaries which were confirmed and messenger – dependent eukaryotic cell-free translation systems that appeared at the end of the decade (Laycock and Hunt, 1969; Lockard and Lingrel, 1969; Mathews and Korner, 1970).

Many mechanisms were built on these foundations which further explain some of the important aspects of translation in both prokaryotes as well as eukaryotes. This emerged in the 1970s. There a question arose which was of initiation site selection and this largely accounted for in prokaryotes by the Shine – Dalgarno sequence base-pairing along with 16S ribosomal RNA (Shine and Dalgarno, 1975).

For the problem in eukaryotes, a solution came later in the period of the 1980s along with the recognition of cap dependent scanning and the Kozak consensus sequence (Kozak and Shatkin, 1978). Further, this was followed by identification of IRES-dependent mechanisms (Jang et al. 1988; Pelletier and Sonenberg 1988).

The optimal codon hypothesis further explained a bias in the usage of synonymous codons. This can influence the production levels of individual genes (Ikemura 1981) and the signal hypothesis which considered the protein transport into the ER that is the endoplasmic reticulum.

The facilitation of the investigations of mRNA translation and the metabolism was done by the discovery in the early period of the 1970s of the terminal hallmarks which was found on most eukaryotic mRNAs 5′ caps (Adams and Cory 1975; Both et al. 1975; Furuichi and Miura 1975; Furuichi et al. 1975a,b; Perry and Kelley 1975; Wei and Moss 1975) and 3′ poly(A) tails (Darnell et al. 1971a,b; Edmonds et al. 1971; Lee et al. 1971).

By the ending period of the 1980s, the identification and purification of most of the components of the translation system and the reconstitution of their activities *in vitro* further resulted in the pathway of protein synthesis and this had been defined. Moreover, this was well understood in outline (Hershey et al. 2018).

There were subsequent detailed analyses that resulted in an in-depth understanding of a number of mechanisms of initiation, elongation and termination in bacteria (Rodnina, 2018). Considering the eukaryotic cells, this was done along with indispensable contributions from yeast genetics as well as biochemistry (Dever et al. 2018; Hellen, 2018; Merrick and Pavitt, 2018). This has given rise to several questions related to regulation that need to be addressed at levels of sophistication which is ever-increasing.

8.4. INSIGHTS TO RIBOSOMAL HISTORY FROM TRNA STRUCTURE

tRNA that is transfer RNA is said to be at the core of the modern translation machinery. Here, it tends to facilitate both the specificity of the process through the codon-anticodon interaction as well as the actual chemistry of peptide bond formation. From now sometimes, the structure of tRNA has been known and now, recently, it has been redetermined at a much higher resolution.

The tRNA molecule contains the essence of two domains while pointing out explicitly and this includes the charging or the synthesis domain containing the CCA stem and the TΨC stem or loop and the codon or anticodon recognition domain. This domain further contains the D stem or loop and the anticodon stem or loop. There have been persuasive arguments that showed that the two domains historically evolved separately and the older among them is the charging/synthesis domain.

Considering such a view, there is the ability to make peptides presumably preceding the role of the ribosome while making the peptide bond synthesis as a template directed process. There were further observations that showed that a direct duplication of an RNA hairpin can lead to a secondary structure which is like a cloverleaf. Such type of duplication event can further result in the formation of the main tertiary interactions which in turn characterize the modern tRNA also.

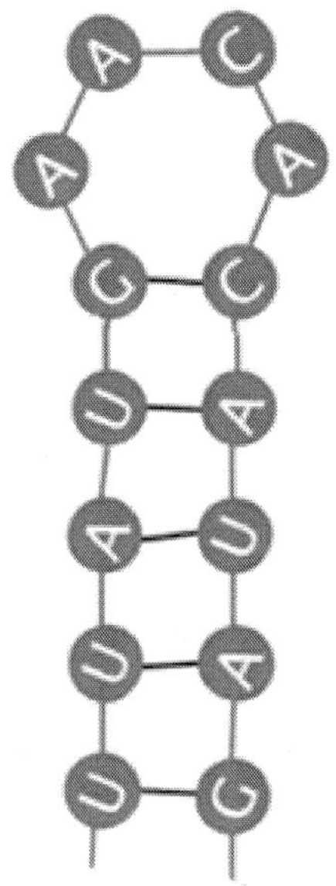

Figure 8.2. RNA Hairpin Loop

Source: Image by Wikimedia Commons

Originally, this was shown for alanine and now in today's scenario, it is well established for a number of tRNAs. It showed that a surprisingly small RNA incorporates the portion of the tRNA which further contains the acceptor stem and sometimes the TΨC stem/loop. This can particularly be charged with the cognate amino acid. The minihelix domain alone recognized by many proteins and enzymes also, for example, the ribozyme- RNAse P, the tRNA nucleotidyl transferase, T54 methyltransferase and EF-Tu and it suggested that this domain can be quite ancient.

Contrastingly, it is not even required to have the anticodon to get recognized by many class II tRNA synthetases. Thus, it further seems plausible that the anticodon domain is not much ancient. And therefore, the structure of tRNA provides us a working hypothesis for the order of early events in translation. The starting of the translation machinery would take place along with a primitive "half tRNA" charging or the synthesis system and this is further followed by a "whole tRNA" charging or template directed synthesis system. Ultimately, this is followed by a whole ribosome mediated translation system. There are fundamental conformational changes that tend to facilitate the movement of mRNA relative to the tRNA. These changes would likely to be presented from the earliest stages. Considering their part, the synthetases have been complex proteins which could not have existed until a later time in the process. The original synthetase was a ribozyme and certain experimental evidence further suggests this is feasible, and this has been widely speculated.

In such stages, there might have been the evolution of the modern synthetases. Initially, they may have facilitated the actions of ribozymes, which are added additional useful capability including the ability to bind the amino acid or affinity for ATP. Eventually, it displaced the ribozyme altogether.

There has been evolution of the ability to recognize and transfer the peptide to the minihelix domain of tRNA later in the period and ultimately, the ability to recognize identity of tRNA (that is especially as defined by the anticodon) last.

8.5. BASIC THEORIES OF THE CODE NATURE, ORIGIN, AND EVOLUTION

Nevertheless, there are many deviations and all of these deviations involve only a few codons. Therefore, considering its main features, it can be said that the structure of the code seems not to have changed through the entire history of life. Precisely in other words, if said to be more, this is since the time of the LUCA that is Last Universal Common Ancestor of all modern (or cellular) life forms. It is important to have explanations about the universality of the genetic code and the manifest non - randomness of its structure.

The frozen accident or code expansion theory of Crick can be said to be a default explanation. This explanation does not require any kind of special mechanisms and such an explanation is only predicted while existing of a LUCA along with an advanced translation system and this resembles the modern one. This is the implicit assumption and that is the LUCA was not a progenote with primitive as well as very inaccurate translation.

However, most of the time, this explanation is considered to be unsatisfactory on the most general as well as epistemological grounds. The reason behind this is, in a sense, a non – explanation and also, because the existence of variant codes and the flexibility of the code (that is revealed through experiments) provides some challenge to the frozen – accident view.

There is a fact that there seem to be ways for sneaking in the changes to the standard code and yet, the evolution of the same limited modifications is seen to have happened independently in various lineages. This in turn suggests that the code structure could be non – accidental.

There are three main theories that are not necessarily mutually exclusive. These theories have been proposed in attempts while attributing the pattern of amino acid assignments in the standard genetic code to physiochemical or biological factors or a combination thereof.

Remarkably, considering each of these theories, it can be said that there are central ideas behind these. These central ideas have been formulated during the classic age of molecular biology. This was not long after the period when the code was deciphered or even much earlier and it remained relevant to this day, although there was number of subsequent developments. The three theories with respect to the historical contexts and below there have been discussions about the current status of each of these theories.

1. There is a stereochemical theory. This theory asserts that specifically for amino acids, the codon assignments are determined by a physicochemical affinity. This affinity exists in between the amino acids and the cognate nucleotide triplets (which include codons or anticodons). Therefore, considering this class of models, it can be said that the specific structure of the code is not at all accidental but instead of this, it is necessary and possibly unique.

The very first stereo-chemical model was developed by Gamow in the year 1954. This was the period which was almost immediately after the resolution of the structure of DNA has been done. This was done effectively along with the idea of the code itself. An explicit mechanism was proposed by Gamow to relate amino acids and holes that were rhomb - shaped and this was formed by various nucleotides in DNA.

Subsequently, after the period when the code was deciphered, there have been more realistic stereochemical models which were proposed. But, generally, these were deemed improbable and the reason was the failure of direct experiments for the identification of specific interactions between amino acids and cognate triplets. However, there is an inherent attractiveness of the stereochemical theory that makes it much easier to see the way in which the evolution of code started further stimulated experimental as well as theoretical activity in this area.

2. There was a second theory and that was an adaptive theory of the code evolution postulates. This theory includes the structure of the genetic code which was shaped under selective forces which made the code maximally robust. This means that the forces minimize the effect of errors on the structure as well as the function of the synthesized proteins.

There has been a possibility that there is a difference between the "lethal-mutation" hypothesis under which the evolution of standard code takes place to minimize the effect of point mutations and the "translation-error minimization" hypothesis. This hypothesis posits the most important pressure in the evolution of code which was the selection for minimizing the effect of the translational mis-readings. When the combination of the two types of forces is used then it is said to be conceivable as well. There is a fact that is related to the codons code for similar amino acids and the experimental observations and that is the occurrence of the mistranslation takes place more frequently in the first and third positions of codons. Whereas, on the other hand, it is the second position that correlates best with amino acid properties and they were construed as evidence in support of the adaptive theory.

In addition to this, the translation error minimization hypothesis has received some statistical support from Monte Carlo simulations and this later became a major tool to analyze the degree of optimization of the standard code.

3. The third theory is named as the co-evolution theory. This theory posits that the structure of the standard code tends to reflect the pathways of amino acid biosynthesis. As per today's scenario, the code has been co-evolved with the amino acid biosynthetic pathways that are done, during the code evolution, for precursor amino acids, the subsets of codons have been re-assigned for encoding the product amino acids.

Although there exists a basic idea related to the co-evolution hypothesis and this is similar in the scenario of code extension of Crick. The explicit identification of precursor - product pairs of amino acids as well as strong statistical support for the inferred precursor - product pairs further gained much attention and acceptance of co-evolution theory.There is a complementary approach considering the problem of code evolution and this approach espouses a "tRNA-centric" view. Under this view, the features of the code are determined by various types of co-evolution including that of the codons and the cognate tRNA anticodons or of the codons and aminoacyl-tRNA synthetases.

There has been the interpretation of the co-evolution and this has been done in terms of minimization of the rate as well as the effect of translation errors or, at the early stages of the code evolution, with respect to the reduction of coding ambiguity.

8.6. THE ORIGIN OF TRNA

The genomic tag hypothesis was proposed by Wiener and Maizels and that is the 3′ (or 'top') half of tRNA majorly functioned as a tag further demarking the 3′-end of genomic RNAs to replicate and therefore, it was considered to be the first part of tRNA to evolve. The phylogenetic evidence has been published by Sun and Caetano-Anollés and their belief supports the genomic tag hypothesis and this was done by confirming, "*that the 'top half' of tRNA is more ancient than the 'bottom half*". According to the observations done by Noller, the tRNA top half which contains the T arm and the acceptor stem, including the amino acid binding site, tends to interact almost exclusively with the large ribosomal subunit, whereas, on the other hand, the bottom half which contains the D and anticodon arms, tends to interact almost exclusively with the small subunit.

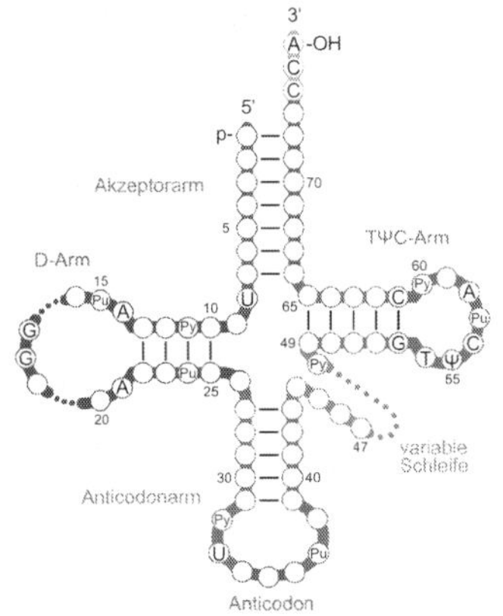

Figure 8.3. The tRNA cloverleaf structure.

Source: Image by Wikimedia Commons

Usually, the peptide synthesis (that is defined as a function of the large subunit) is viewed as more ancestral while compared with decoding (which is defined as a function of the small subunit). This was the reason why a view, which has been supported by a structural analysis by Bokov and Steinberg (2009) – the top half of tRNA (that tends to interact with the large subunit), has been considered to be more ancestral in comparison to the bottom half.

Nevertheless, there has been a conflict in this 'standard model' for the origin of tRNA, and the results of Sun and Caetano-Anollés which provide its support to this model. This conflict has been with more recent findings of Harish and Caetano Anollés (2012) that was on the relative ages of the ribosomal subunits.

As it has been described above, there have been findings which further suggest that the small ribosomal subunit was the first to evolve and this is difficult to reconcile along with the fact that the bottom half of tRNA (along with which the interaction of the small subunit mainly takes place), is, by theirs and others estimation, the newer half of tRNA.

In an equal manner, there was their finding that the large ribosomal subunit evolved more recently and this finding is difficult to reconcile. This was done along with the fact that the top half of tRNA (along with which the interaction of the large subunit mainly takes place), is the older half of tRNA and this was by the estimations done by them and others.

There was a finding of Caetano-Anollés and colleagues (2008) related to the most ancient tRNAs that coded for selenocysteine, serine, tyrosine, and leucine. Incidentally, this finding runs counter to other work in the area but also, appears to have contradictions with their finding that says " *the variable region was the last structural addition to the molecular repertoire of evolving tRNA substructures*".

8.7. SEQUENCE-BASED EVOLUTIONARY ANALYSIS OF THE AARSS

There has been evolutionary analysis which has been sequence-based. This sequence-based evolutionary analysis of the AARSs has considered being the subject of intense study. The documentation of the evolution of these proteins has been done at around three levels of divergence that include specificity, subclass and class. The specificity level involves the studies which tend to focus on the evolution of individual AARSs particularly for some specified amino acid.

Each and every class has been synthesized and roughly grouped into the standard subclasses. The subclass level tends to examine the series of gene duplications which arises a subset of the AARSs of each class. This subset is thought to be more closely in relation to each other in comparison to other AARSs of the same class.

There is a most divergent level which includes the events related to gene duplication. These events gave rise to all of the synthetases in each class individually. The quality as well as quantity of the investigations into the AARSs further decreases along with the increasing level of divergence. There have been the most accurate work done which involves the specificity level and slightly beyond this level to some extent.

There have been sequence comparison between AARSs of different specificities and this typically occupies a range of sequence identity values that tend to fall below the twilight zone threshold that is less than 25 percent sequence identity. The sequence comparison is quickly considered to be unreliable at this and lower levels of sequence identity. Considering this regime of similarity, it becomes more difficult to provide the difference between correctly aligned homologous sequences and sequences or random alignments which are unrelated.

There is a most recent as well as insightful analysis of the synthetases at the specificity level and this analysis is found in the paper of Woese (2000). In a contemporaneous paper, Wolf also examined this level of AARS evolution in a detailed manner.

Even though, the evolutionary analysis of the catalytic domain which has been presented is more complete and extensive, the paper by Wolf (1999) tends to provide an invaluable discussion of the accessory domains (which include anticodon binding and other domains) that are common to subsets of the synthetases within and in some of the cases across both classes. Considering most of the part, the results of these papers are said to be consistent with respect to the catalytic domain.

There has been phylogenetic analysis of every AARSs separately. Considering the three domains of life, this analysis further shows the distribution of each AARS. The categorization of the pattern of this distribution can be done while keeping in mind the universal phylogenetic tree and can be classified as the basal canonical, full canonical or non - canonical pattern.

Class I synthetases conform to the full canonical pattern and these are specific for W, Y, E, L, and I. whereas, on the flipped side, the class II synthetases are specific for H, D, F, and P. The basal canonical pattern can be observed for the class I synthetases which further specify M, V, and R. Class II synthetases follow the basal pattern that is specific for A and T.

As per the reports presented by Woese, it can be said that the class I LysRS represents the non - canonical pattern. But there is a more recent study which included more sequences and it concluded that this synthetase exhibits the basal pattern. Taking into account each of the synthetases which have been mentioned above, some fit the canonical or basal canonical pattern, but all exhibit exceptions to these patterns which are said to be the result of different levels of horizontal gene transfer.

On one hand, LeuRS conforms completely to the canonical pattern without exception, whereas on the other hand, for ProRS, a number of bacteria have received the archaeal type through HGT. There are the synthetases that exhibit the basal canonical pattern and considering these, HGT from Bacteria to the eukaryotes accounts for the majority of the deviations from the full canonical pattern. This is the case for ThrRS, ValRS, and AlaRS.

There are some bacterial MetRSs that are distinct of the archaeal genre. These tend to have received the gene through HGT from Archaea. Moreover, the eukaryotic MetRSs are considered to be of the archaeal genre, but they fail while forming a distinct grouping that is required for the full canonical pattern to hold. There are several instances in which the sequence analysis can go beyond constructing phylogenies for the specificities alone and show some of the gene duplication events that result in the formation of the AARS specificities.

Because of the close homology between AspRS and AsnRS and a close homology between GluRS and GlnRS, Woese could complete the evolutionary path which further gave rise to AsnRSand GlnRS. From the eukaryotic GluRS, the evolution of GlnRS takes place through gene duplication and after that, it was horizontally transferred to a limited number of Bacteria.

Somewhat earlier, there was an evolution of AsnRS as the result of gene duplication, before the advent of the Eucarya, from the ancestral archaeal type AspRS. In such type of cases, the differences between the archaeal genre and the bacterial genre for AspRS (GluRS) are greater in comparison to the differences between the archaeal type AspRS (GluRS) and AsnRS (GlnRS).

There was a sequence-based analysis which was done by Brown and Doolittle (1995). This sequence-based analysis depicts the duplication events that further gave rise to LeuRS, IleRS, and ValRS and demonstrates that these synthetases group monophyletically taking into account the amino acid specificity. The sequence comparison also represents that TrpRS and TyrRS group monophyletically.

Considering a greater level of divergence, the structure to guide the alignment of subclass IIA AARSs was used by Schimmel and coworkers (2001). Along with the structural alignment of ProRS, ThrRS, SerRS, GlyRS, HisRS, and AspRS as a template, they further aligned the most conserved sequence fragments of all of the available sequences for these synthetases.

The AspRS was used to root the class IIA phylogeny. There were results thus obtained that indicated that SerRS, ProRS, and ThrRS tend to form a distinct supercluster but that the exact relationships between these three synthetases could not be determined. GlyRS and HisRS were each reported to form groups separately.

Although this phylogeny was guided by the alignment of regions which are structurally equivalent across the subclass IIA, the construction of the resulting phylogeny was done by the maximum parsimony method. This method is strictly depend on sequence. Considering then an earlier study, Nagel and Doolittle (1991) used a number of alignments of highly conserved sequence fragments of the AARSs to construct a phylogeny along with a sequence-based similarity measure, at the class level.

Even though only a limited set of AARS sequences, which include no archaeal examples, was said to be available, this early work provided suggestions that remain accurate when the more rigorous study and larger data sets were put into considerations. There were further observations provided by Nagel and that included that the AARSs cluster monophyletically taking into account the amino acid specificity, and the monophyly rule remains intact while having only two exceptions.

The phylogeny further represents a number of superclusters which appear to be consistent or can be partially consistent along with the data and some other more recent investigations. These superclusters are indicated by parentheses such as for class I (GluRS, GlnRS), (TrpRS, TyrRS), and (ValRS, IleRs, LeuRS, MetRS); for class II (ThrRS, ProRS, SerRS) and (LysRS, AspRS, AsnRS).

Since the sequence identity in between a number of the AARSs having different specificities is below the twilight zone threshold, the connections within and among these superclusters are not so reliable and can be misleading in some of the cases.

8.8. AN "RNA-FIRST" VIEW OF THE ORIGIN OF LIFE

8.8.1. Abiotic Synthesis of Polynucleotides

The particular section would consider oligonucleotides synthesis from ß-D-nucleoside 50 phosphates, as of now excluding the query as to how nucleotides started existing on the primitive Earth. The chemical reactions involved are of basically of two types. First, the conversion of the nucleotide into an activated derivative would be required. For example, a nucleoside 5′-polyphosphate.

Another would be, a reaction between oligonucleotide molecule or 3' hydroxyl group of nucleotide and the activated monomer phosphate group is ought to be executed. Discussions pertaining to oligonucleotides synthesis from 30 phosphates nucleotides would not be taking place. It is because reaction of activated nucleoside 20 or 30 phosphates would usually occur easily, forming 20, 30 cyclic phosphates.

It would be improbable for the cyclic phosphates to effectively oligomerize as 1.0 L/mol would be the sole sequence acts as the equilibrium constant belonging to dimer formation (Erman and Hammes 1966; Mohr and Thach 1969). The complementary template present would result in the formation of rather larger oligomers, driving forward the reaction of chain extension due to the free energy associated with the chain extension.

The nucleoside 50 triphosphates (NTPs) would be the substrates of polymerization in the case of enzymatic RNA and DNA synthesis. Although considered as a degradative natural enzyme, yet, polynucleotide phosphorylase could be used in the process of synthesizing oligonucleotides from nucleoside 50 diphosphates. Therefore, the apparent match intended for the activated nucleotide forms would be the nucleoside 50 polyphosphates.

The synthesis of nucleoside 50 tetraphosphates belonging to nucleotides and inorganic trimetaphosphate would deliver a practically probable prebiotic route towards activated nucleotides, even though the formation of nucleoside 50 triphosphates would not occur easily (Lohrmann, 1975). Reports pertaining to more or less probable alternate prebiotic syntheses of nucleoside 50 polyphosphates belonging to nucleotides have been available (Handschuh et al. 1973; Osterberg et al. 1973; Reimann and Zubay 1999)

Mobilizing of the first phosphate for converting the nucleosides to 50 nucleotides, would be, however, not very evident. The nucleoside 50 polyphosphates would be considered as high energy phosphate esters, but in case of an aqueous solution, they would be comparatively unreactive. For enzyme catalyzed polymerization it might be considered beneficial, but for the nonenzymatic polymerization of nucleoside 50 polyphosphates, it would be a great problem. Further, it would be occurring at a much slower pace than the hydrolysis of the subsequent polynucleotide.

A per a different tactic related to the nucleotides activation, an activated intermediate would be refrained from being isolated through the usage of a carbodiimide, a condensing agent (Khorana, 1961). Though it is considered to be a prevalent method in organic synthesis, yet its usage would be troublesome in prebiotic chemistry. Cyanamide and cyanoacetylene would be considered as the potentially prebiotic molecules and supposed to be activating the nucleotides in aqueous solution.

However, the following condensation reactions would be incompetent (Lohrmann and Orgel, 1973). Maximum attempts for studying nonenzymatic polymerization of nucleotides related to prebiotic chemistry would have been applying nucleoside 50 phosphoramidates, specifically nucleoside 50 phosphorimidazolides.

The formation of phosphorimidazolides could occur from imidazole and nucleoside 50 polyphosphates (Lohrmann, 1977), nevertheless, as prebiotic molecules, their probability would be just marginal. Since they could be prepared simply as well as react at a suitable rate in aqueous solution, hence, they were selected. There would be three primary nucleophilic groups in nucleotides: the 50 phosphate, the 20 hydroxyl and the 30 hydroxyl group, as per the decreasing reactivity order.

Thus, nucleotide or oligonucleotide reaction with an activated nucleotide would generally comprise of 50, 50 pyrophosphate, 20, 50 phosphodiester and 30, 50 phosphodiester linked adducts, in a decreasing abundance order (Sulston et al. 1968). Hence, by various monomers condensation would possibly comprise an oligomer with one pyrophosphate and a 20, 50 phosphodiester linkages preponderance. The scope to produce complete 30, 50 linked oligomers from activated nucleotides would be less till the time a catalyst could be located, increasing the proportion of 30, 50 phosphodiester linkages.

Many metal ions, Pb^{2+} and UO_2^{2+} specifically, would catalyze the oligomers formed from nucleoside 50 phosphorimidazolides (Sleeper and Orgel 1979; Sawai et al. 1988). While being performed in an eutectic solution of the activated monomers (in concentrated solutions obtained through more dilute solutions being partially freeze) the Pb2+ catalyzed reaction is considered to be significantly efficient. The eutectic conditions would result in the formation of the considerable amount of long oligomers, however, product oligomers would have a large quantity of 20, 50 linkages constantly (Kanavarioti et al. 2001; Monnard et al. 2003).

8.8.2. Non – enzymatic Replication of RNA

If there would have been a mechanism for polymerization of activated nucleotides on the primitive Earth, a compound mixture of product oligonucleotides differing in length and order both would have been generated. The following stage associated with the emergence of an RNA World is considered to have been the imitation of few of the said molecules, to commence an equivalent natural selection process.

Template -directed synthesis would be a vital reaction in the replication of nucleotides. That would imply synthesizing of a complementary oligonucleotide directed by a pre-existing oligonucleotide. In regards to the aspect of nonenzymatic replication, a considerable amount of work has already been performed. Since the work is supposed to have been reviewed in some other place (Joyce 1987; Orgel 2004a), hence, a summarized result would be provided here.

Among all the other conclusions provided, the first primary one concludes that a greater number of activated nucleotides would not experience effective, regiospecific, template directed reactions with an RNA or DNA template being present. Usually, a small part of template molecules would be able to direct the synthesis of an entire complement and that would usually comprise of a mixture of 20, 50 and 30, 50 phosphodiester linkages.

An extensive search resulted in locating a set of activated nucleotides experiencing effective and highly regiospecific template directed reactions. The possibility of poly(C) being able to direct the synthesis of long oligo(G)s in a highly effective and highly regiospecific reaction was exhibited while working with guanosine 50 phospho-2-methylimidazolide (2-MeImpG). Incubation of poly(C) with an equimolar mixture related to the four 2-MeImpNs (N ¼ G, A, C, or U), the product comprising of noncomplementary nucleotides would be less than 1% (Inoue and Orgel 1982).

As suggested by successive experiments, the above mentioned and other reactions in the relation which would have been discussed later would preferably occur within the double helices context with the A form RNA resembling structure (Kurz et al. 1997, 1998; Kozlov et al. 1999, 2000). An additional C residue would be contained within the random copolymers and used for direct synthesis of products with G and complements pertaining to alternate bases existing in the template (Inoue and Orgel 1983).

A poly(C, G) template reaction would be regarded exclusively regarded interesting due to the composition of the products, like the template, which would completely include C and G residues. Such products being used as templates, likewise, might result in being emerged as a self-replicating sequence. However, self-replication would be improbable, primarily because poly(C,G) molecules which would not have an extra C residue would and tend to form constant self-structures preventing them from acting as templates (Joyce and Orgel 1986).

There would be two types of self-structures:

- The standard Watson-Crick type that would be based on C-G pairs
- A Quadra helix structure resulting from the association of four G-rich sequences

Consequently, whichever C nucleotide could be competent as a suitable template would give rise to G-rich complementary products, usually being locked in self-structure, hence, unable to act as templates.

It would be quite difficult to overcome the self-structure issue by the usage of standard C and G nucleotides. That's because it is supposed to necessitate the discovery of conditions favoring the mononucleotides bond for allowing template-directed synthesis to occur, but suppressing the long duplex regions to be formed excluding activated monomers from the template.

Moreover, in the field designated as "synthetic biology" is considered to have been witnessing a lot of efforts for the achievement of nonenzymatic replication with molecules though similar to the biological nucleic acid not controlled by the considerations of probable prebiotic chemistry. For instance, 20 and 30 hydroxyl groups related to activated mononucleotides could be substituted by an amino group in whichever position. It would be able to provide improved nucleophilicity and would result in extra rapid template-dependent (as well as template-independent) oligomerization (Lohrmann and Orgel 1976; Zielinski and Orgel 1985).

Dinucleotide building blocks with 30 amino and 30 deoxynucleotide links could even be oligomerized through a good condensing agent (Zielinski and Orgel 1987). Further modifying of the nucleotide bases would be helpful in executing the template-directed replication of nucleic acid sequence constituting all four bases (Schrum et al. 2009). Though such efforts would not be able to comprehend the origin of the RNA World, it would contribute in the interpretation of the chemical challenges that are supposed to be conquered for the achievement of the nonenzymatic RNA replication.

8.9. THE FIRST RNA REPLICASE

The notion of the RNA World tends to emphasize on an RNA molecule which further catalyzes its own replication. Such a molecule is required to function as an RNA – dependent RNA polymerase which tends to act on itself in order to produce complementary RNAs and acting on the complementary RNAs to produce additional copies of itself. The efficiency as well as fidelity of this process is required to be sufficient for producing the viable 'progeny' RNA molecules at a rate which tends to exceed the rate of decomposition of the parents.

The details of the replication process go beyond these requirements which are not highly constrained. The RNA is said to be the first view of the origin of life and this assumes that there is an availability activated ß-D-nucleotides by some that are yet an unrecognized abiotic process.

Moreover, it assumes that there is the existence of a means to convert the activated nucleotides to an ensemble of random – sequence polynucleotides. This is said to be a subset of which had the ability to replicate. It seems to be implicit in the model that such polynucleotides further replicate themselves but do not replicate unrelated neighbors and so can be whatever reason.

It is still unclear, whether the replication involves one molecule copying itself (and its complement) or a family of molecules which copy each other altogether. These are the questions which are set aside for the moment to consider firstly, the question of whether an RNA molecule of reasonably short length can tend to catalyze its replication with sufficiently high fidelity.

8.9.1. Replicase Function in the Evolved RNA World

Although it is difficult to say the way in which the first RNA replicase ribozyme arose, it is moreover not difficult to imagine the way in which a molecule would function once it is developed. The involvement of the template – directed polymerization of mono- or short oligonucleotides is there in the chemistry of RNA replication while using the chemistry in a number of ways which is similar to that used by the contemporary group I ribozymes (Cech 1986; Been and Cech 1988; Doudna and Szostak 1989).

There is one important difference and that is an RNA replicase would more likely make use of a different leaving group and this is not like the group I ribozymes which rely on a nucleoside or oligonucleotide leaving group. This, in turn, provides a substantial driving force for polymerization. One more difference is that after its release, it does not get to involve in some competing phosphoester transfer reaction.

8.9.2. From Ligases to Polymerases.

The polymerization of activated nucleotides tends to proceed through a nucleophilic attack by 30 -hydroxyl of an oligonucleotide that is the template - bound at the a-phosphorus of an adjacent template-bound nucleotide derivative. The nucleotide is said to be activated to attack by the presence of a phophoryl substituent examples of which include a phosphate, polyphosphate, alkoxide, or imidazole group. There are polyphosphates such as inorganic pyrophosphate which are the most obvious candidates for the leaving group.The condensation reaction tends to assist by favorable orientation of the reactive groups, stabilization of the trigonal-bipyramidal transition state, deprotonation of the nucleophilic 30 -hydroxyl and charge neutralization of the leaving group. These all are the tasks that are performed by RNA (Narlikar and Herschlag 1997; Emilsson et al. 2003). These tasks can either act alone (Ortoleva-Donnelly and Strobel 1999) or along with the suitably positioned metal cation or other co-factor (Shan et al. 1999; Shan et al. 2001).

There is a possibility and that there can be the existence of an RNA replicase ribozyme that has been made abundantly clear by work further including the ribozymes which have been further developed in the laboratory by *in vitro* evolution (Bartel and Szostak 1993; Ekland et al. 1995; Ekland and Bartel 1996; Robertson and Ellington 1999; Jaeger et al. 1999; Rogers and Joyce 2001; Johnston et al. 2001; McGinness and Joyce 2002; Ikawa et al. 2004; Fujita et al. 2009).

Let's take an example. Bartel and Szostak (1993) started with a large population of random – sequence RNAs and evolved the class I RNA ligase ribozyme. This was said to be an optimized version of which is about 100 nucleotides in length and it catalyzes the joining of two template – bound oligonucleotides.

The process of condensation tends to occur between the 30 -hydroxyl of one oligonucleotide and the 50 -triphosphate of another. This, in turn, forms a 30, 50 -phosphodiester linkage and release inorganic pyrophosphate. This reaction is classified as ligation and the reason is the nature of the oligonucleotide substrates, but also, that it involves the same chemical transformation as it is catalyzed by modern RNA polymerase enzymes.

8.9.3. RNA Replication

Although it falls short of the ultimate goal of a general-purpose RNA polymerase ribozyme, for RNA-catalyzed, a robust reaction system uses RNA replication and this has been shown. A pair of cross-replicating ligase ribozymes are used by the system that each catalyzes the formation of the other by further using a mixture of four different substrate oligonucleotides (Lincoln and Joyce, 2009).

There are reaction mixtures which contain only these RNA substrates such as $MgCl_2$, and buffer. In such reaction mixtures, a small starting amount of ribozymes is required in order to give rise to a number of additional ribozymes by using a process of RNA - catalyzed exponential amplification. Whenever there is depletion in the substrates, the replication process can be started again and sustained indefinitely by replenishing the supply of substrates.

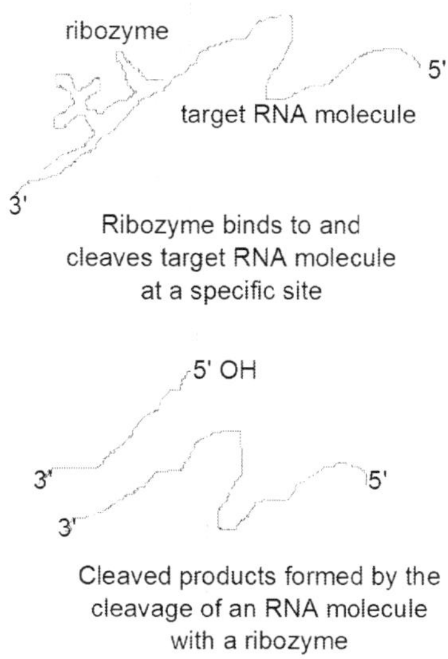

Figure 8.4. Ribozyme mechanism

Source: Image by Wikipedia

Through specific Watson-Crick pairing interactions, the recognition of the substrates can be done by the ribozymes. Along with this, the evolution experiments can be performed while providing a variety of substrates that have a number of different sequences in these recognition regions and different corresponding sequences in the catalytic domain of the ribozyme. RNA replication was performed along with a library of 144 possible substrate combinations. This results in the emergence of a set of replicators which is highly advantageous. Also, it included recombinants which were absent at the beginning of the experiment. Until and unless, the advent of a general-purpose RNA polymerase ribozyme takes place, the best platform is provided but the system of cross-replicating ligases to study the biochemical properties as well as evolutionary behavior of an all - RNA replicative system.

8.9.4. Nucleotide Biosynthesis

Probably, the RNA replicase activity is not considered to be the only catalytic behavior that was required for the existence of the RNA World. It has been made the top priority to maintain an adequate supply of the four activated nucleotides. Despite the fact that a large reservoir of these compounds was there in the prebiotic environment, the reservoir would ultimately become depleted, and there can be a requirement of some capacity for nucleotide biosynthesis. The ribozymes have been obtained, and this is done by *in vitro* evolution. This process tends to catalyze some of the steps of nucleotide biosynthesis.

8.10. CONCLUSION

As per the view of RNA World Hypothesis, protein translation (especially considering the origin as well as the evolution of the ribosome) shepherded the transition from the primordial biota further depending on RNA to the modern world. When the high conservation of the catalytic core of the ribosome is done then it tends to raise fundamental questions related to the origin and evolution of the translation system. Furthermore, important experimental insights are currently missing which have been provided by understanding the structural and substrate flexibility of the ribosome.The conclusion that can be drawn from the above chapter is that the new foundational technologies for the biosynthesis of sequence - defined polymers from engineered translation machinery further provide a transformative toolbox for synthetic and chemical biology. It is just similar to organic chemistry once revolutionized the ability of chemists to build molecules which were done by following a basic set of design rules.

REFERENCES

1. Bernhardt, H., 2012. The RNA world hypothesis: the worst theory of the early evolution of life (except for all the others)a. *Biology Direct*, [online] 7(1), p.23. Available at: <https://www.researchgate.net/publication/229089462_The_RNA_world_hypothesis_The_worst_theory_of_the_early_evolution_of_life_except_for_all_the_others> [Accessed 9 May 2020].
2. Burton, Z. and Lei, L., 2020. *Evolution Of Life On Earth: Trna, Aminoacyl-Trna Synthetases And The Genetic Code*. [ebook] Available at: <https://www.mdpi.com/2075-1729/10/3/21/pdf> [Accessed 9 May 2020].
3. Fox, G. and Naik, A., n.d. The Evolutionary History of the Translation Machinery. *The Genetic Code and the Origin of Life*, [online] pp.92-105. Available at: <https://www.researchgate.net/publication/285333100_The_Evolutionary_History_of_the_Translation_Machinery> [Accessed 9 May 2020].
4. Fox, G., Tran, Q., Rivas, M. and Stepanov, V., 2013. 3 Origins and evolution of the translation machinery. *Journal of Biomolecular Structure and Dynamics*, [online] 31(sup1), pp.2-2. Available at: <https://www.researchgate.net/publication/263682030_Origins_and_evolution_of_the_translation_machinery> [Accessed 9 May 2020].
5. Hernandez, G., 2012. On the Emergence and Evolution of the Eukaryotic Translation Apparatus. *Cell-Free Protein Synthesis*, [online] Available at: <https://www.intechopen.com/books/cell-free-protein-synthesis/on-the-emergence-and-evolution-of-the-eukaryotic-translation-apparatus> [Accessed 9 May 2020].
6. Koonin, E. and Novozhilov, A., 2009. *Origin And Evolution Of The Genetic Code: The Universal Enigma*. [ebook] Available at: <https://iubmb.onlinelibrary.wiley.com/doi/pdf/10.1002/iub.146> [Accessed 9 May 2020].
7. Lu, D., Lu, T. and wu, H., 2012. *Origin Of Life, RNA World Or Alternate RNA World*. [online] Available at: <https://www.researchgate.net/publication/268280254_Origin_of_life_RNA_world_or_alternate_RNA_world> [Accessed 9 May 2020].
8. O'Donoghue, P. and Schulten, Z., 2003. *On The Evolution Of Structure In Aminoacyl-Trna Synthetases*. [ebook] Available at: <https://mmbr.asm.org/content/mmbr/67/4/550.full.pdf> [Accessed 9 May 2020].

9. Robertson, M. and Joyce, G., 2010. The Origins of the RNA World. *Cold Spring Harbor Perspectives in Biology*, [online] 4(5), pp.a003608-a003608. Available at: <https://www.pmf.unizg.hr/_download/repository/Origins_of_the_RNA_World%5B1%5D.pdf> [Accessed 9 May 2020].
10. Adams and Cory 1975
11. Anderson et al., 1983
12. Bartel and Szostak 1993
13. Bartel and Szostak, 1993
14. Been and Cech 1988
15. Both et al. 1975
16. Brenner et al. 1961
17. Carlson et al., 2012
18. Cech 1986
19. Chapple et al. 2003
20. Chen and Zubay, 1983
21. Craig et al., 1992
22. Darnell et al. 1971
23. Dever et al. 2018
24. Doudna and Szostak 1989
25. Edmonds et al. 1971
26. Ekland and Bartel 1996
27. Ekland et al. 1995
28. Emilsson et al. 2003
29. Erickson and Blobel, 1983
30. Fujita et al. 2009
31. Furuichi and Miura 1975
32. Furuichi et al. 1975
33. Ganoza et al., 1985
34. Gesteland, 1966
35. Gros et al. 1961
36. Handschuh et al. 1973
37. Hellen 2018

38. Ikawa et al. 2004
39. Inoue and Orgel 1982
40. Inoue and Orgel 1983
41. Jackson and Hunt, 1983
42. Jaeger et al. 1999
43. Jang et al. 1988
44. Jewett and Swartz, 2004a
45. Jewett et al., 2008
46. Johnston et al. 2001
47. Joyce 1987
48. Joyce and Orgel 1986
49. Kanavarioti et al. 2001
50. Khorana 1961
51. Kim et al., 2006
52. Kozak and Shatkin 1978
53. Kozlov et al. 1999, 2000
54. Krieg and Melton, 1987
55. Kung et al., 1977
56. Kurz et al. 1997, 1998
57. Lamborg and Zamecnik, 1960
58. Lau and Unrau 2009
59. Lau et al. 2004
60. Laycock and Hunt 1969
61. Lee et al. 1971
62. Lincoln and Joyce, 2009
63. Liu et al., 2005
64. Lockard and Lingrel 1969
65. Lohrmann 1975
66. Madin et al., 2000
67. Mathews and Korner 1970
68. Matthaei and Nirenberg, 1961
69. McGinness and Joyce 2002
70. Merrick and Pavitt 2018

71. Michel-Reydellet et al., 2004
72. Mikami et al., 2006b
73. Monnard et al. 2003
74. Narlikar and Herschlag 1997
75. Nevin and Pratt, 1991
76. Nirenberg and Matthaei, 1961
77. Nirenberg and Matthaei, 1961
78. Orgel 2004a
79. Osterberg et al. 1973
80. Pelham and Jackson, 1976
81. Pelletier and Sonenberg 1988
82. Perry and Kelley 1975
83. Pratt, 1984
84. Reimann and Zubay 1999
85. Roberts and Paterson, 1973
86. Robertson and Ellington 1999
87. Rogers and Joyce 2001
88. Ryabova et al., 1998
89. Sawai et al. 1988
90. Sawasaki et al., 2007
91. Schrum et al. 2009
92. Schweet et al., 1958
93. Shan et al. 2001
94. Shimizu et al., 2001
95. Shimizu et al., 2005
96. Shine and Dalgarno 1975
97. Shirokov et al., 2007
98. Sleeper and Orgel 1979
99. Spirin et al., 1988
100. Unrau and Bartel, 1998
101. Wei and Moss 1975
102. Yang et al., 1980
103. Zamecnik et al., 1948

104. Zawada and Swartz, 2006
105. Zhou et al., 2012
106. Zielinski and Orgel 1985
107. Zubay, 1973
108. Zubay, 1973

INDEX

N

O

P